T0331578

Constructed Wetlands

Constructed Wetlands
Hydraulic Design

Saeid Eslamian

Saeid Okhravi

Faezeh Eslamian

CRC Press
Taylor & Francis Group
Boca Raton London New York

CRC Press is an imprint of the
Taylor & Francis Group, an **informa** business

CRC Press
Taylor & Francis Group
6000 Broken Sound Parkway NW, Suite 300
Boca Raton, FL 33487-2742

© 2020 by Taylor & Francis Group, LLC
CRC Press is an imprint of Taylor & Francis Group, an Informa business

No claim to original U.S. Government works

International Standard Book Number-13: 978-0-367-19689-9 (Hardback)
978-0-429-24262-5 (eBook)

Library of Congress Cataloging-in-Publication Data

Names: Eslamian, Saeid, author. | Okhravi, Saeid, author. | Eslamian, Faezeh A., author.
Title: Constructed wetlands : hydraulic design / authored by Saeid Eslamian, Saeid Okhravi and Faezeh Eslamian.
Description: Boca Raton : CRC Press, [2020] | Includes bibliographical references.
Identifiers: LCCN 2019038689 (print) | LCCN 2019038690 (ebook) | ISBN 9780367196899 (hardback) | ISBN 9780429242625 (ebook)
Subjects: LCSH: Constructed wetlands--Design and construction. | Environmental hydraulics.
Classification: LCC TD756.5 .E75 2020 (print) | LCC TD756.5 (ebook) | DDC 628.3/5--dc23
LC record available at https://lccn.loc.gov/2019038689
LC ebook record available at https://lccn.loc.gov/2019038690

Visit the Taylor & Francis Web site at
http://www.taylorandfrancis.com

and the CRC Press Web site at
http://www.crcpress.com

Contents

Preface

THE STATE OF CONSTRUCTED WETLAND (CW) science and technology can be obtained thorough exploration of papers and experience related to the theoretical, experimental and numerical development and application of this technology. CW hydraulic characteristics, such as residence time, mixing, and short-circuiting directly control treatment performance and efficiency of chemical reactions within each CW. This book focuses on the research aspects that involve a comprehensive understanding of the hydraulic functioning of CWs. It begins with a brief introduction of CWs followed by different applicable types related to flow direction. Then, hydraulic theory background on CWs is described considering classical ideal and non-ideal flows, and their characterization by such hydrodynamic parameters as plug-flow ratio, short-circuiting, dead-volume ratio, dispersion number, and hydraulic efficiency. The main method by which wetland scientists and engineers gain information about hydraulic processes is through use of inert tracers that enable tracking of water movement through a CW. This book reviews and summarizes the main theory and techniques applied when implementing hydraulic tracer tests in CW treatment systems to obtain the water retention-time distribution across the wetland, with special view paid to the practical issues and needs to be considered through planning, designing and operation phases.

The book also describes the CW design from the hydraulics perspective as related to treatment performance; including the main design parameters such as aspect ratio, substrate porous media type and size, flow inlet/outlet arrangements, CW water levels, and hydraulic loading rates. The performance of CWs can be enhanced by modifying such parameters resulting in hydraulic flow behavior that approaches that of an ideal flow system. This book is intended primarily as a textbook at the graduate/research level and as a guide for field engineers, enabling them to remain current with scientific developments. The final chapter suggests future trends for CW researchers to increase the treatment efficiency of these engineered systems. Therefore, as a simple prerequisite, the readers should have basic background knowledge in hydraulics and a general understanding of environmental processes.

Authors

Saeid Eslamian is a full professor of environmental hydrology and water resources engineering in the Department of Water Engineering at Isfahan University of Technology, where he has been since 1995. His research focuses mainly on statistical and environmental hydrology in a changing climate. In recent years, he has worked on modeling natural hazards, including floods, severe storms, wind, drought, pollution, water reuses, sustainable development and resiliency, etc. Formerly, he was a visiting professor at Princeton University, New Jersey, and the University of ETH Zurich, Switzerland. On the research side, he started a research partnership in 2014 with McGill University, Canada. He has contributed to more than 600 publications in journals, books, and technical reports. He is the founder and chief editor of both the *International Journal of Hydrology Science and Technology* (IJHST) and the *Journal of Flood Engineering* (JFE). Eslamian is now associate editor of three important publications: *Journal of Hydrology* (Elsevier), *Eco-Hydrology and Hydrobiology* (Elsevier), *Journal of Water Reuse and Desalination* (IWA), and *Journal of the Saudi Society of Agricultural Sciences* (Elsevier).

Professor Eslamian is the author of approximately 35 books and 180 chapter books.

Dr. Eslamian's professional experience includes membership on editorial boards, and he is a reviewer of approximately 100 Web of Science (ISI) journals, including the *ASCE Journal of Hydrologic Engineering, ASCE Journal of Water Resources Planning and Management, ASCE Journal of Irrigation and Drainage Engineering, Advances in Water Resources, Groundwater, Hydrological Processes, Hydrological Sciences Journal, Global Planetary Changes, Water Resources Management, Water Science and Technology, Eco-Hydrology, Journal of American Water Resources Association, American Water Works Association Journal*, etc. UNESCO has also nominated him for a special issue of the *Eco-Hydrology and Hydrobiology Journal* in 2015.

Professor Eslamian was selected as an outstanding reviewer for the *Journal of Hydrologic Engineering* in 2009 and received the EWRI/ASCE Visiting International Fellowship in Rhode Island (2010). He was also awarded outstanding prizes from the Iranian Hydraulics Association in 2005 and Iranian Petroleum and Oil Industry in 2011. Professor Eslamian has been chosen as a distinguished researcher of Isfahan University of Technology (IUT) and Isfahan Province in 2012 and 2014, respectively. In 2016, he was a candidate for national distinguished researcher in Iran.

He has also been the referee of many international organizations and universities. Some examples include the U.S. Civilian Research and Development Foundation (USCRDF), the Swiss Network for International Studies, the Majesty Research Trust Fund of Sultan Qaboos University of Oman, the Royal Jordanian Geography Center College, and the Research Department of Swinburne University of Technology of Australia. He is also a member of the following associations: American Society of Civil Engineers (ASCE), International Association of Hydrologic Science (IAHS), World Conservation Union (IUCN), GC Network for Drylands Research and Development (NDRD), International Association for Urban Climate (IAUC), International Society for Agricultural

Meteorology (ISAM), Association of Water and Environment Modeling (AWEM), International Hydrological Association (STAHS), and UK Drought National Center (UKDNC).

Professor Eslamian finished Hakimsanaei High School in Isfahan in 1979. After the Islamic Revolution, he was admitted to IUT for a BS in water engineering and graduated in 1986. After graduation, he was offered a scholarship for a master's degree program at Tarbiat Modares University, Tehran. He finished his studies in hydrology and water resources engineering in 1989. In 1991, he was awarded a scholarship for a PhD in civil engineering at the University of New South Wales, Australia. His supervisor was Professor David H. Pilgrim, who encouraged him to work on "Regional Flood Frequency Analysis Using a New Region of Influence Approach." He earned a PhD in 1995 and returned to his home country and IUT. In 2001, he was promoted to associate professor and in 2014 to full professor. For the past 24 years, he has been nominated for different positions at IUT, including university president consultant, faculty deputy of education, and head of department. Eslamian is now director for center of excellence in Risk Management and Natural Hazards (RiMaNaH).

Professor Eslamian has made three scientific visits to the United States, Switzerland, and Canada in 2006, 2008, and 2015, respectively. In the first, he was offered the position of visiting professor by Princeton University and worked jointly with Professor Eric F. Wood at the School of Engineering and Applied Sciences for one year. The outcome was a contribution in hydrological and agricultural drought interaction knowledge by developing multivariate L-moments between soil moisture and low flows for northeastern U.S. streams.

Recently, Professor Eslamian has published the editorship of nine handbooks published by Taylor & Francis (CRC Press): the three-volume *Handbook of Engineering Hydrology* in 2014, *Urban Water Reuse Handbook* in 2016, *Underground Aqueducts Handbook* (2017), the three-volume *Handbook of Drought and Water Scarcity* (2017), *Constructed Wetlands: Hydraulic Design*

(2020). *An Evaluation of Groundwater Storage Potentials in a Semiarid Climate* by Nova Science Publishers is also his joint book publication in 2019.

Saeid Okhravi is a PhD candidate of hydraulic engineering and research assistant at Bu-Ali Sina University (BASU), Hamadan, Iran. He is currently working on the numerical simulation of flow–structure–sediment interactions on bridge pier scour with the aid of Faculty of Engineering of University of Porto (FEUP), Porto, Portugal. He earned both his bachelor's degree in water engineering specializing in rainwater harvesting systems and a master degree in hydraulic engineering with a specialization in investigation on hydraulic behavior and treatment efficiency of a constructed wetland from Isfahan University of Technology (IUT), Iran.

Saeid was selected as the best PhD student among the students of water engineering department of Bu-Ali Sina University for three years in a row, 2016–2018. Also in 2018, he was awarded outstanding prizes as one of the excellent PhD students based on educational-research perspective from both Ministry of Science, Research and Technology and Iran's National Elites Foundation and received the Visiting International scholarship in FEUP, Porto, Portugal.

Mr. Okhravi is the author of one Persian book, five book chapters, and more than 40 scientific publications and technical reports. He has contributed to the *Handbook of Engineering Hydrology* (Vol: Fundamentals and Applications, Title: *Groundwater-Surface Water Interactions*), *Urban Water Reuse Handbook* (Title: *Water Reuse in Rainwater Harvesting*, Title: *Urban Water Reuse: Future Policies and Outlooks*), *Handbook of Drought and Water Scarcity*

(Vol: Environmental Impacts and Analysis of Drought and Water Scarcity, Title: *Water Conservation Techniques,* Title: *Drought in Lake Urmia*) by Taylor & Francis Group (CRC Press).

Faezeh Eslamian is a PhD holder of Bioresource Engineering in McGill University. Her research focuses on the development of a novel lime-based product to mitigate phosphorus loss from agricultural fields. Faezeh completed her bachelor's and master's degrees in Civil and Environmental Engineering from Isfahan University of Technology, Iran, where she evaluated natural and low-cost absorbents for the removal of pollutants such as textile dyes and heavy metals. Furthermore, she has conducted research on the worldwide water quality standards and wastewater reuse guidelines. Faezeh is an experienced multidisciplinary researcher with interest in soil and water quality, environmental remediation, water reuse, and drought management.

CHAPTER 1

An Introduction to Constructed Wetlands

Saeid Eslamian, Saeid Okhravi, and Mark E. Grismer

1.1 GENERAL

In this fast-changing and highly interconnected world, problems related to water availability, scarcity, quality, and the associated conflicts are numerous, complicated, and challenging. Efforts to effectively resolve these problems require a clear vision of future water availability and demand as well as new ways of thinking, developing, and implementing water planning and management practices. Reclaiming and reusing treated wastewater can create an alternate water source for secondary use by reducing demand on potable water sources utilized for drinking water (Eslamian 2016; Eslamian et al. 2016).

Nowadays, the use of constructed wetlands (CWs) for urban stormwater and wastewater treatment is widely adopted in many urban environments, many of which are successfully incorporated into urban landscapes (Okhravi et al. 2016; Jabali et al. 2017). First, as compared to conventional energy-intensive treatment technologies (physical–chemical–biological treatments), CWs are an attractive and stable alternative due to low cost and energy savings (Zhang et al. 2009). Second, CWs can provide potentially valuable wildlife habitat in urban and suburban areas (Rousseau et al. 2008), as well as an esthetic value within the local natural environment. Finally, CWs are often beneficial in small- to medium-sized towns due to easy operation and maintenance, providing a useful complement to the traditional sewage systems used predominantly in larger cities. Thus, the introduction of CWs in municipal wastewater treatment systems has been something of a revolution. The incorporation of CWs aids in future design plans for new wastewater treatment facilities and provides a greater understanding of CW benefit for residents (Vymazal 2019). Abundant research indicates that wetlands can filter wastewater in an environmentally and economically efficient manner. With their low mechanical input and high environmental benefits, development of CWs appears to be a good tool for the water resources planners considering wastewater treatment and reuse.

Like conventional wastewater treatment plants (WWTPs), CWs are engineered systems, but are much lower-cost construction, operation, and maintenance alternative for wastewater treatment and reclamation where applicable. They deploy eco-technological biological wastewater treatment technology, have an excellent pollutant removal performance, and enjoy lower energy consumption. CW systems are designed to mimic processes found in natural wetland ecosystems. Thus, CWs are an environmentally friendly technology that helps to reduce environmental impacts associated with wastewater treatment by treating waste on-site with low energy and chemical consumption (Flores et al. 2019).

CWs are shallow water bodies typically planted with water-tolerant vegetation common to the locale. The basin substrate material is usually sand or gravel that enables root penetration and sedimentation filtration and pollutant uptake processes to remove pollutants from wastewater. A CW is comprised of five primary design aspects including basin dimensions, substrate materials, type and density of vegetation, and liner and inlet/outlet configuration. Generally, wastewater enters the CW basin through a distribution manifold and flows over the surface and/or through the substrate and discharges from the basin through a structure that controls the CW water depth (Sudarsan et al. 2015).

CWs reduce contaminant concentrations through a complex of physical, chemical, and biological mechanisms that occur through water, substrate, plants, and microorganism interactions (Chang et al. 2012). Particular removal mechanisms depend on the wastewater constituents, type of substrate, and available plant/microbial species in the system. Physical procedures are mainly filtration by the substrate layer, plant-root zone, and sedimentation. The main chemical reactions consist of chemical precipitation, adsorption, cation exchange, and oxidation/reduction reactions (Ramond et al. 2012). It is widely acknowledged that biochemical reactions attributed to bacterial communities (biofilms) play the most important role in pollutant transformation in aerobic and anaerobic conditions (Chang et al. 2012; Samsó and García 2013).

Design considerations include hydrodynamic factors aimed at optimizing performance of the various physical, chemical, and biological treatment processes within the CW system. The macrophytes in CWs can support a number of very important pollutant removal mechanism, including removal of suspended solids and associated contaminants (e.g., nutrients, metals, organic contaminants, and hydrocarbons). In addition, design considerations for stormwater treatment CWs include the botanical structure and layout of the wetland and the hydrologic regime necessary to sustain the botanical structure. However, the operating conditions of these systems are stochastic, with intermittent and highly variable hydraulic and pollutant loading. The achievable treatment by a CW system can be improved by careful consideration of such factors as the CW shape, vegetation, wetland volume, and the hydrologic and hydraulic effectiveness. The relationship between residence time and pollutant removal efficiency is largely influenced by the targeted particulate settling velocity and reaction kinetics that are controlled by the CW hydraulics. To date, previous publications on CW systems have focused on comprehending the processes leading to the pollutant removal. Research on flow hydraulics through the CW porous media (substrate) has concentrated primarily on the assessment of relationships and interactions between microbial communities and plants and the reduction of pollutants in the system (Steer et al. 2002; García et al. 2005; Caselles-Osorio et al. 2007). However, there are areas of CW internal hydraulic functioning that remain poorly understood. Comparatively, there is less research dedicated to CW hydraulic design. Thus, this book assesses and elaborates CW hydraulic design aspects to obtain the best CW hydraulic and treatment performance (Persson et al. 1999; Zahraeifard and Deng 2011; Alcocer et al. 2012; Wang et al. 2014a, 2014b; Okhravi et al. 2017). Our objective with this book is to provide information about the internal hydraulic behavior of CWs as described by hydraulic parameters like hydraulic efficiency, active/dead zones distribution, and short-circuiting as related to pollutant removal efficiency.

1.2 CW CONFIGURATIONS

Wetlands can be effective in removing biochemical oxygen demand (BOD), total suspended solids (TSSs), nitrogen, and phosphorus, while also reducing the metals, organics, and pathogen content of storm- or wastewaters (Kadlec and Wallace 2008). CW design configurations are classified according to the flow pattern through the CW system, type of wastewater treated, basin substrate, dominant macrophytes, operational usage type, and desired level of treatment. Among these criteria, there are three main CW types considered in the literature based on flow patterns: free-water surface (FWS), horizontal subsurface flow (HSSF), and vertical subsurface flow (VSSF) systems (IWA 2000).

1.2.1 Free-Water Surface CWs

FWS CWs are defined by the exposure of the water surface to the atmosphere and shallow ponded depths (~0.3 m) where treatment is achieved largely within the water column (EPA 2000). SSF wetlands (natural and constructed) are defined by most flows below the substrate (e.g., sand/gravel) surface (Kadlec 2009). Both wetland types may contain a variety of plants including submerged, emergent, and floating plants. Many FWS wetlands have similar characteristics to natural marshes that provide additional wildlife habitat benefit (Figure 1.1).

FWS CWs tolerate variable water levels and nutrient loads and can achieve a high removal of suspended solids and BOD and moderate removal of pathogens, nutrients, and other pollutants, such as heavy metals, depending on the CW residence time and ambient temperatures. This type of wetland is more appropriate for low-strength wastewater following primary or secondary treatment sufficient to reduce wastewater BOD due to limited aeration (Vymazal 2011). FWS CWs are a good option where land is readily available and/or inexpensive, and they work well

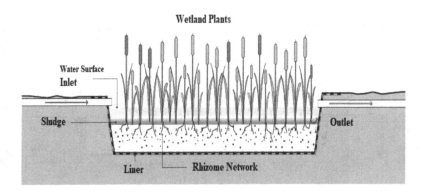

FIGURE 1.1 Schematic of the FWS wetlands. (Adapted from Tilley, E. et al., *Compendium of Sanitation Systems and Technologies*, 2nd Rev. ed., Swiss Federal Institute of Aquatic Science and Technology (Eawag), Duebendorf, Switzerland, 2014.)

in warm climates. The hydraulic efficiency of FWS CWs depends, in part, on how well the wastewater is distributed at the CW inlet (Tilley et al. 2014). Wastewater typically flows into the CW using weirs, orifices, or a distribution manifold pipe.

1.2.2 Subsurface Flow (SSF) CWs

Greater wastewater treatment (at comparatively lower flow rates than in FWS CWs) is possible in SSF CWs (Vymazal 2011). These systems rely on water flow through the substrate material in a horizontal (HSSF) or vertical configuration (VSSF). HSSF CWs are typically a shallow basin (depths of 0.3–1.0 m) filled with filter material (usually sand or gravel substrate) and planted with vegetation tolerant of saturated or submerged conditions (Figure 1.2). Water levels within the SSF CW are maintained at several centimeters below the substrate surface through control of outlet conditions. Depending on the wastewater inlet–outlet conditions, the wastewater moves through a wetland substrate more-or-less horizontally. Within the wetland substrate, the wastewater encounters a network of aerobic, anoxic, and anaerobic zones associated with relative depth, plant-rooting depths, and overall

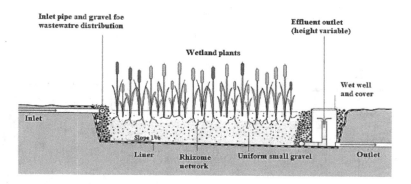

FIGURE 1.2 Schematic of the HSSF wetlands. (Adapted from Tilley, E. et al., *Compendium of Sanitation Systems and Technologies*, 2nd Rev. ed., Swiss Federal Institute of Aquatic Science and Technology (Eawag), Duebendorf, Switzerland, 2014.)

CW depth. Aerobic zones occur near the substrate surface and around the plant roots and rhizomes that leak oxygen into the substrate (UN-HABITAT 2008). However, much of the saturated bed is anaerobic under most wastewater loadings. Wastewater treatment generally occurs by filtration adsorption and microbiological degradation processes. Typical plant species include common reed (*Phragmites australis*), cattail (*Typha* spp.), and bulrush (*Schoenoplectus* spp.). HSSF CWs can effectively remove wastewater organic pollutants (TSS,[1] BOD, and COD[2]) as well as various nutrients, metals while breaking down some industrial or pharmaceutical chemicals.

VSSF CWs are also a planted filter domain for secondary or tertiary treatment of wastewater (e.g., gray- or black-water) that enters at substrate surface, gradually percolates, and drains at the bottom (Figure 1.3). As possible with other systems, intermittent dosing (4 to 10 times a day) of the VSSF CWs enable

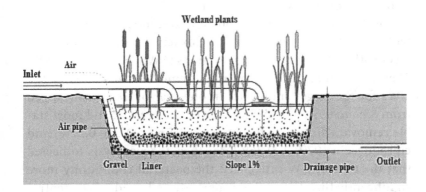

FIGURE 1.3 Schematic of the VSSF wetlands. (Adapted from Tilley, E. et al., *Compendium of Sanitation Systems and Technologies*, 2nd Rev. ed., Swiss Federal Institute of Aquatic Science and Technology (Eawag), Duebendorf, Switzerland, 2014.)

---

[1] Total Suspended Solids.
[2] Chemical Oxygen Demand.

the substrate to oscillate between aerobic and anaerobic conditions, thereby facilitating nitrification (Tilley et al. 2014). The oxygen diffusion from the air created by the intermittent dosing system contributes much more to the filtration bed oxygenation as compared to oxygen transfer through the plant (Brix 1997; UN-HABITAT 2008). However, the primary role of vegetation is to maintain permeability in the filter and provide a habitat for microorganisms. Schematic illustrations of horizontal and vertical SSF CWs are shown in Figures 1.2 and 1.3, respectively.

Generally, some type of primary wastewater treatment is essential to limit clogging within SSF CW substrate and ensure effective treatment. Depending on the situation, such a pretreatment might include septic tanks, sand or rotating screen filters, and anaerobic baffle reactors (Hoffmann and Platzer 2010). A well-designed inlet layout could improve the removal efficiency of these types of CWs by preventing short-circuiting.

By way of an example, Liu et al. (2019) investigated the removal efficiencies and accumulated concentrations of the antibiotics oxytetracycline and ciprofloxacin in three pilot-scale CWs with different flow configurations (FWS, HSSF, and VSSF). They found that the average mass removal efficiencies of the CWs for the target antibiotics in one year of treatment ranged from 85% to 99%. The VSSF achieved the greatest and most stable removal efficiencies for oxytetracycline (99% ± 0.27%) and ciprofloxacin (97% ± 0.26%). Further, their analysis indicated that the type of CW influenced the bacterial community more significantly than the wetland substrate and residual concentration of antibiotics.

1.2.3 Hybrid CWs

Greater nitrogen removal from domestic wastewater was one of the primary drivers to develop CW technology. The nitrogen degradation process includes nitrification of ammonia under aerobic conditions followed by denitrification of nitrate under anaerobic conditions at the same time. That is why single stage of vertical or horizontal CW designs without intermittent loading remains unsatisfactory, since both aerobic and anaerobic conditions cannot be developed at the same time. However, depending on the wastewater loading rates, type of manifold distribution at the inlet and load cycling, SSF CWs can be operated with both aerobic and anaerobic zones (Grismer et al. 2001). While aerobic conditions facilitate nitrification due to great oxygen transport ability by the intermittent feeding regime, they also increase sludge production that results in the clogging of the substrate material as sometimes seen in VSSF CWs. Besides, denitrification is the prevailing nitrogen-transforming process in HSSF wetlands under the anaerobic conditions provided by a regime of permanent saturation when sufficient carbon is available (Ávila et al. 2017). Hence, combination of VSSF and HSSF CWs has become a common strategy to improve the total nitrogen removal and reduce risks of substrate clogging. The hybrid system could be either HSSF wetland followed by VSSF wetland or VSSF wetland followed by HSSF wetland as in the sand prefilter VSSF followed by the much larger HSSF CW employed by Shepherd et al. (2001).

More recently, use of zeolite aggregate substrate combined with culturing anammox bacteria within the substrate biofilms has enabled removal of BOD, suspended solids, and nitrogen from a range of partially treated wastewater strengths (Grismer and Collison 2017; Collison and Grismer 2018a, 2018b). However, the focus of this book is on the hydrodynamic design of the CWs and the most important factors that affect internal hydraulic behavior through the wetland cells.

SUMMARY

Constructed wetlands have long been applied to improve stormwater or wastewater quality. These systems have proved their effective treatment of BOD, TSS, nitrogen, and phosphorus, as well as for reducing metals, organics, and pathogens.

The three basic types of constructed wetlands have been discussed as FWS, HSSF, and VSSF wetlands. In those systems, the wastewater will pass through a network of aerobic, anoxic, and anaerobic zones. The aerobic zones take place at the vicinity of roots and rhizomes, which leak oxygen into the substrate. Major removal mechanisms and treatment performance have been reviewed especially well in this chapter.

The designers are confronted with the wetland type selection, along with size, layout, and many other decisions, and the best choice may require expert evaluation. It was the purpose of this chapter to present information from a large data set to assist the designer in evaluating the main decision elements. It should be noted that some overarching concepts like cost and areal extent of the system are determining factors.

CHAPTER 2

Hydraulic Theory

Constructed Wetland

Saeid Okhravi, Saeid Eslamian,
and Mark E. Grismer

2.1 GENERAL

Pollutant removal within CWs occurs through a wide range of interactions between the suspended sediments, substrate, microorganisms, litter, plants, atmosphere, and wastewater as it moves through the CW system. The nature and dynamics of water movement through the wetland can have a significant influence on the extent of these interactions. Many of the important biogeochemical reactions rely on contact time between wastewater constituents and microorganisms and/or substrate, while wastewater velocity and temperature are important determining factors for other pollutant removal processes, such as sedimentation (Headley and Kadlec 2007). Flow short-circuiting or dead zones within wetland cells reduce contact time that effectively increases flow velocities and adversely impacts treatment efficiency.

The hydraulic performance of CWs, in terms of removal efficiency, depends on many factors among which the most significant ones are basin volume, hydraulic loading rate (HLR), and hydraulic retention time (HRT). The hydraulic retention time (HRT), a critical parameter in designing CWs, is the average time required for water parcels to move from the inlet to the outlet. The nominal (or theoretical) HRT (nHRT) is defined as:

$$\text{nHRT (days)} = t_n = \frac{V_n}{Q} = \frac{\phi(l \times w \times h)_n}{Q} \tag{2.1}$$

where V_n is the wetland water volume (m^3) and Q is the flowrate (m^3/day), ideally taken to be the average of the inflow and outflow rates, l is the wetland length (m), w is the wetland width (m), h is the wetland depth (m), and ϕ is the substrate porosity (dimensionless).

The actual retention time of water within the wetland may be shorter than the nominal one due to the occurrence of dead zones, or in some cases longer due to inaccuracies in flowrate and/or wetland volume measurements. Climatic effects, such as rainfall and evapotranspiration can also cause the actual residence time in a wetland to vary dynamically. Furthermore, due to various degrees of mixing, dispersion, and hydraulic inefficiencies, wetlands have a range, or distribution, of retention times (Kadlec 2000). That is, water parcels that enter a wetland at time zero do not all leave simultaneously after the computed time, but will leave the wetland after varying lengths of time, both shorter and longer than t_n. This distribution of times that various fractions of water spend in a wetland is termed as the residence time distribution (RTD). The hydraulic performance in wetlands can be analyzed by studying water flow patterns or HRT distributions (RTD).

Kadlec and Knight (1996) provide a comprehensive review of RTD theory, with specific reference to CWs. Levenspiel (1972), Fogler (1992), and Grismer et al. (2001) provide literature reviews about water RTDs as determined using chemical reactor theory as applied to wetland cells and their measurement using nominally inert tracers. The following section is a summary of the principal elements of this theory.

2.2 IDEAL AND NONIDEAL FLOW

There are two flow reactor types that form the basis of steady-state flow theories in CWs that depend on the degree of mixing within the reactor or wetland. Ideal plug-flow within a reactor is characterized by wastewater flows with no water mixing or diffusion along the flow path. In this way, all water parcels have equal reaction or flow time within the system. Under steady-state flows, a plug-flow reactor (PFR) has a single residence time (equal to t_n) that is equivalent to the actual residence time. Early designs of CW assumed plug-flow conditions (Crites 1988; Reed 1990; Reed et al. 1995), whereas more recent designs are based on the continuously stirred tank reactor (CSTR) concept (Werner and Kadlec 2000). In this concept, fluid entering the reactor is instantaneously mixed with reactor contents, while the fluid exiting the CSTR has the same composition as the fluid within the reactor (Levenspiel 1972). In this way, all parcels of wastewater have equal probability of leaving the wetland at a given moment (Kadlec 1994). Under steady-state flow conditions, a CSTR experiences a range of residence times, and an exponential decay characterizes the distribution of residence times in the reactor (Headley and Kadlec 2007). Another variation of the CSTR model is the tanks-in-series (TIS) model. In this model, the wetland is partitioned into a number of equal-sized CSTRs and the concentration, C, of a pollutant leaving each "tank" is equal to the uniform internal concentration (Kadlec and Knight 1996). The number of tanks, N, in the TIS model represents the degree of mixing. A high value of N means a small degree of dispersion and flow conditions within the reactor approach that of a PFR (Persson and Wittgren 2003).

Wastewater flows through a CW are typically nonideal, reflecting a combination of the processes outlined above. Importantly, the distribution of residence times may depend not only on mixing processes but may result from vertically stratified or horizontally distributed differences in flow velocities (Kadlec 2000; Grismer et al. 2001). For example, in FWS wetlands, vertical and horizontal variations in water velocity can be caused by spatial patterns in the topography of the bed and vegetation density. In addition, water moves more slowly through the plants, and more rapidly in surface waters in unobstructed channels (Kadlec and Knight 1996; Eslamian et al. 2014). In SSF CWs, velocity profiles can result from spatial variability in substrate permeability, root biomass, and accumulated clogging solids (Grismer et al. 2001). An extreme case of clogging or overloading the SSF CW occurs when wastewater flows over the substrate surface (short-circuiting). The effective design of inlet and outlet structures and the shape and bathymetry of the wetland can also be important (Persson et al. 1999). Wetlands with single-point inlet and outlet configurations are prone to short-circuiting as water follows the shortest or most transmissive flow path between the inlet and outlet. Ideally, the inflow should be uniformly distributed across the entire inlet cross-sectional surface area, while the outflow is collected across the entire outlet width of the wetland. Although irregular wetland shapes may be more aesthetically appealing, they are generally hydraulically less efficient and prone to the development of dead zones and backwater areas when compared to rectangular wetlands. Flows with a uniform velocity profile are approached only in FWS with a large L:W aspect ratio (Persson 2000) as eddy and recirculation flows may otherwise occur. Ultimately, all of these variations in flow velocities cause some water parcels to move more quickly to the wetland outlet than others (Werner and Kadlec 1996). Therefore, in practice, local velocity profiles or velocity vectors provide limited information for the evaluation of CW hydraulic performance as compared to the net effects of various velocity variations across the CW provided by the bell-shaped RTD curve.

2.2.1 Tracer Study

Flow patterns across and through CWs are rarely equivalent to simple PFR or CSTR conditions, and field tracer studies are required to evaluate how the water actually moves through a particular CW (Grismer et al. 2001; Suliman et al. 2005). The tracer is assumed to follow the same flow pattern as the water parcel with which it entered the wetland and should, therefore, give a reasonable reflection of the hydraulic RTD (Headley and Kadlec 2007). The resultant tracer RTD can then be used to elucidate the actual wetland water volume that is involved in treatment (hydraulic efficiency), as well as the degree of apparent mixing and deviation from ideal flow (Werner and Kadlec 1996).

Tracer studies deploy a relatively inert chemical as a tracer that is readily measured, has low toxicity, has high solubility, is at low or nonexistent background concentrations (Chang et al. 2012), and is inexpensive (e.g., Cl or Br salts). The tracer is added at the CW inflow structure as an "impulse" type injection or "step" to continuously added with the wastewater. Sampling for the tracer at the CW outlet and preferably at several sampling transects along/across the CW provides valuable information about the water flow paths within the CW and its hydraulic performance. The choice of impulse or continuous addition of the tracer at the CW inlet depends on the analytical model chosen for analysis; however, in both, the effluent tracer concentrations are measured as a function of time, C(t), and the hydraulic parameters of interest are determined. The function C(t) depends on tracer velocities and dispersion within the porous matrix (Clark 1996) and can be identified anywhere that C(t) is measured. The residence time distribution (RTD), f(t), is obtained through normalizing C(t) by the total tracer mass injected, or that measured at a particular location. Of course, other hydraulic parameters such as porosity ($\varnothing$), hydraulic conductivity (K), dispersivity, short-circuiting, and effective volumes can also be calculated indirectly through other analyses.

For example, using the method of moments, hydraulic parameters can be easily derived from f(t) (Jury et al. 1991; Clark 1996). The moments, M_N, are defined as

$$f(t) = \frac{C(t)}{\displaystyle\int_0^\infty C(t)\,dt}$$

$$M_N = \int_0^\infty t^N f(t)\,dt \tag{2.2}$$

where $M_0 = 1.0$ by definition; the first moment, M_1, is the mean retention time (i.e., mean residence time, $mHRT = t_m = t_{mean}$), the average time that a tracer particle spends in the water system is also defined as the centroid of the RTD, where the RTD function, $f(t)$, is represented by the concentration or mass in Equation 2.3. In this equation, C is the tracer concentration, Q is the flow discharge, and, t is corresponding time for each tracer concentration, which can be derived from the RTD curve by using tracer studies.

$$t_m = \frac{\int_0^\infty tCQdt}{\int_0^\infty CQdt} \qquad (2.3)$$

In SSF CWs, t_m is often less than the plug-flow detention time, t_n, because of dead-end flow zones, or otherwise poor hydraulic pathways (such as corner zones). The mean tracer velocity or flow velocity, u, at measurement location from the inlet l (flow length), is given by the ratio l / t_m. The variance, σ^2, given by Equation 2.4, is a measure of the spread of the RTD curve; a variance equal to zero implies ideal plug-flow condition (i.e., no dispersion other than the advection). The σ^2 is used to determine the effective dispersion coefficient, d, at corresponding distance, l, followed by Equation 2.5.

$$\sigma^2 = \frac{\int_0^\infty (t_m - t)^2 f(t)dt}{\int_0^\infty f(t)dt} \qquad (2.4)$$

$$d = \frac{l^2}{2} \frac{\sigma^2}{t_m^3} \qquad (2.5)$$

The Peclet number, P_e, at distance l along the tank is given by ul/d. Kadlec and Knight (1996) use the dispersion number (D), $1/P_e$, to characterize the significance of dispersive transport. The dispersion number is typically in the range of 0.2–0.4 (Kadlec et al. 1991; Kadlec 2000). Finally, it is also possible to determine t_m and d through the least-squares fitting of tracer RTDs to one-dimensional analytical solutions of the convective-dispersion equation (CDE) such as that given by Jury and Sposito (1985). CSTR and PFR reactor models were often used to describe treatment reactor hydraulics (Kadlec and Knight 2006). Because of variable hydraulic characteristics, CWs may exhibit nonideal flow patterns; outflow RTDs can be fit to a combination of these models. One approach is the tanks-in-series model, which for N tanks and impulse tracer input the outlet tracer concentration function, C(t), is given by

$$C_N(t) = \frac{NC_0}{(N-1)!}\left(\frac{Nt}{t_n}\right)^{N-1} e^{\left(\frac{-Nt}{t_n}\right)} \qquad (2.6)$$

where C_0 is the initial tracer concentration and t_n is the theoretical retention time. Another common measure of the degree of plug flow is the number of stirred tanks (N) used in a tank-in-series model (Fogler 1992; Grismer et al. 2001). In general, the greater the value of N required to fit C(t), the more closely the system approaches plug-flow hydraulic conditions. Values of N can be determined directly from the variance and nominal residence time as t_n^2/σ^2.

The median retention time (often referred to as t_{50}) can in some cases be used as an approximation to the mean retention time. This is correct when the RTD is symmetrical (as in Gaussian functions), but it is less suitable for more commonly occurring skewed RTD curves.

Figure 2.1 displays typical tracer input and ideal response (RTD) curves following step and impulse additions of tracer. Most CW tracer studies use the impulse addition technique

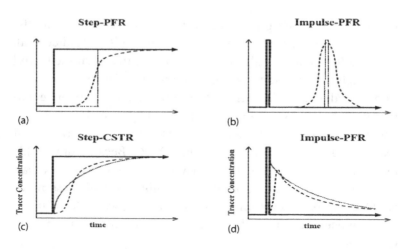

FIGURE 2.1 RTD for ideal (dotted) and nonideal (dashed) plug-flow (a and b) and complete mix (c and d) reactors following step and impulse additions of a tracer at the inlet (solid line). (Adapted from Headley, T.R. and Kadlec, R.H., *Ecohydrol. Hydrobiol.*, 7, 269–282, 2007.)

because it requires far less tracer than a step, or continuous input, and is thus less costly, particularly for large-scale wetland systems. Moreover, use of an impulse injection of a known mass of tracer enables more precise determination of tracer recovery at the CW outlet and further understanding of the wetland hydraulics. Here, we focus on the techniques relevant to the impulse input approach. Impulse tracer additions to CWs typically result in positively skewed bell-shaped exit distributions, with some tracer exiting at short times, and some exiting at longer times.

Practical issues to consider when conducting a hydraulic tracer study include the type and quantity of tracer used, the method of introducing the tracer into the wetland, sampling approaches, and data requirements. Tracer mass recovery is a critical factor in evaluating the relative success of the tracer study and should always be reported. It is generally considered acceptable if at least 80% of the mass of tracer added as an impulse at the inlet is recovered at the outlet (Headley and Kadlec 2007).

The three most popular and practical choices for hydraulic tracers are isotopes, ions, and dyes. Isotope technology has been used to determine the fate and transport of isotopic nitrogen in nutrient treatment. Kadlec et al. (2005) used the stable isotope ^{15}N introduced as ammonium in two subsurface flow (SSF) wetlands receiving primary meat processing water and found little of the tracer in gas emissions (1%). They found that the majority of the tracer were in plants (6%–48%) and sediments (28%–37%). Ronkanen and Kløve (2007, 2008) used the stable isotope $^{18}O/^{16}O$ ratio and tracer tests to prevent or diminish short-circuiting and dead zones in peatlands. Although isotope technology has high accuracy, it is expensive. Radioactive tracers, such as tritium, have very good tracer properties and may be suitable for use in CWs. However, the use of radioactive substances in wetlands is often precluded by strict regulations and specific analytical requirements. Conservative biotracers, such as coliphage MS2, and the bacteriophage of Enterobacter cloacae show promise for use as hydraulic tracers, particularly where there are concerns regarding the toxicity of the various chemical tracers (Hodgson et al. 2003).

Ionic compounds have been widely used as groundwater tracers (Wang et al. 2008). Bromide and lithium are the most extensively used ionic tracers, mainly due to their relatively low cost and ease of analysis. They are typically added as solutions of sodium or potassium bromide, or lithium chloride and have yielded reliable results in numerous wetland studies (Grismer et al. 2001; Smith et al. 2005). Małoszewski et al. (2006) used instantaneously injected bromide to evaluate hydraulic characteristics for a duckweed pond in Mniów, Poland, and found that the wastewater flowed along three different flow paths to the outlet. As bromide, or chloride concentrations in natural waters may be present well above detection, large quantities of ionic tracers may be required to surpass the background levels, thereby making them most suitable for use in small to moderate-sized CWs. Dyes have advantages of low detection, zero natural background, and low relative cost. However, they are susceptible to a variety of environmental influences that can affect their stability and detection (Sabatini 2000). There are some common dyes to select as a tracer, such as Rhodamine, Uranine, and Eosine, with a given concentration within a short time. These tracers can return more than 95% of the injected amount in submerged aquatic vegetation-dominated mesocosms (Dierberg and DeBusk 2005). One of the most suitable dyes is Rhodamine WT because it exhibits the fewest matrix artifacts. Chang et al. (2012) used Rhodamine WT in a subsurface up-flow wetland to explore the interface between hydraulic and environmental performance in concert with a transport model to collectively provide hydraulic retention time (i.e., 7.1 days) and compelling evidence of pollutant fate and transport processes. Research findings indicated that pollution-control media demonstrate smooth nutrient removal efficiencies across different sampling port locations.

One of the other applicable dyes is Na-fluorescein (Uranine). Because of its conservative behavior, low detection limits, and low costs as compared to Rhodamine, Uranine is a widely applied dye for tracer experiments. Naurath et al. (2011) used Uranine in mine water as tracer tests with an extremely low concentrations detection limit (3 ppt). Okhravi et al. (2017) applied Uranine by pulse injection to determine the HRT and the hydraulic patterns in a HSSF CW. They recovered about 90% of the tracer in most cases. The only limitation is that Uranine is best detected under in alkaline conditions (pH ~ 9) and at the standard excitation wavelength (490 nm) (Naurath et al. 2011). Because the pH value has a major influence on the Uranine signal and wastewater pH in most CWs is near neutrality (7–7.5), buffering of the samples to a suitable pH range is inevitable. In all dye tracers, the dye grab samples should be collected in amber or dark glass bottles and kept shaded prior to analysis to limit photolysis and temperature variations.

As a rule of thumb, 30–40 sample points are normally adequate to define the response curve (RTD). Sampling frequency and interval can be determined based on inlet/outlet flow rate or at regular times. With tracer concentration measurements completed in the field in real time, the sampling frequency can be adjusted to adapt to the changing concentrations; that is, more frequent samplings as concentrations begin to rapidly increase or decrease so as to better characterize the tracer RTD. In situations where the flow rate is somewhat variable (the majority of cases), flow-weighted sampling will generally ensure that accurate sampling of the tracer response curve is achieved (Headley and Kadlec 2007). In any case, it is necessary to record both the time of sampling and volume of water that has passed through the sample point since the tracer impulse was added, as these will be required in interpreting the tracer response data. Ideally, continuous measurements of the flow rate entering and exiting the wetland should be collected over the duration of the tracer study to better enable the determination of the tracer recovery fraction. Further, it is important to obtain a reasonably accurate measurement of the CW water volume to later obtain a reasonable normalized RTD calculation of tracer recovery and mean HRT (Werner and Kadlec 2000). An accurate understanding of the changes in wetland volume over time will be particularly important in cases where flow and water volume are nonsteady, such as with wetlands receiving event-driven runoff such as stormwater. The estimated volume also forms the basis for computation of the hydraulic efficiency of the wetland.

2.3 HYDRAULIC ASSESSMENT

While hydraulic tracers are used to derive information about the RTD and hydraulic efficiency of CWs, they can also be used as an experimental tool to reveal useful information about the internal hydrodynamics of CWs (e_v, λ, S, P_e, etc.) and to evaluate the effect of different design variables on flow processes (Grismer 2005).

To know about ideal plug-flow conditions, the difference between the mean retention time, t_m, and the nominal retention time, t_n, is the first speculation. Therefore, Thackston et al. (1987)

suggested the relationship, Equation 2.7, between t_n and t_m as the effective volume ratio (e_v):

$$e_v = \frac{t_m}{t_n} = \frac{V_{effective}}{V_{total}} \tag{2.7}$$

By comparing the theoretical PFR and observed HRTs, certain hydraulic characteristics can be assessed, for example

$$mHRT = t_m > nHRT = t_n :$$

when the flows follow a preferential path and the first peak is seen at an earlier stage of the curve than in the theoretical curve, which peaks at a later stage (Chazarenc et al. 2003).

$$mHRT = t_m < nHRT = t_n :$$

Wastewater that stagnates in the system (i.e., "dead" zones) has limited opportunity for biochemical reactions and reduces the effective wetland (pore) volume.

Persson et al. (1999) proposed the hydraulic efficiency index (λ) and found that it was influenced by the wetland aspect ratio, the relative position of the inlet and the outlet, and any obstruction through water flow lines (see Equation 2.8). The maximum tracer concentration occurs at time t_{peak} (t_p).

$$\lambda = \frac{t_p}{t_n} \tag{2.8}$$

Persson et al. (1999) proposed a simple classification system for determining CW hydraulic performance based on λ values; good ($\lambda > 0.75$), satisfactory ($0.5 < \lambda < 0.75$), and poor efficiency ($\lambda < 0.5$). At hydraulic efficiency >0.7, the CW system has uniform hydraulic conveyance, thus reducing the risk of short-circuiting.

Hydraulic short-circuiting (Lloyd et al. 2003) is clearly important to wastewater treatment and can be caused by many factors, including system geometry, flow rate, and inlet and outlet configurations (Safieddine 2009). As a measure of short-circuiting, the quotient S (dimensionless) is used, where t_{16} is divided by t_{50} (Equation 2.9); here, t_{16} is the time for the passage of the 16th percentile of the tracer through the outlet and t_{50} is the time for the passage of the 50th percentile of the tracer (Persson 2000).

$$S = \frac{t_{16}}{t_{50}} \tag{2.9}$$

Short-circuiting is a common problem in which some influent constituents exit the CW before the necessary treatment (residence) time, resulting in overall poor treatment performance as that associated with CWs having small S values. Multiple flow paths can be manipulated by controlling the areal hydraulic loading rate, $HLR = Q_d/A_S$, where Q_d is the average flow rate through the system (m^3/d) and A_S is the surface area of the system (m^2). Reducing short-circuiting can be very effective toward improving hydraulic efficiency and the flow uniformity needed to achieve the highest PFR and treatment.

Dispersion, D, a measure of the degree of mixing in a wetland is another important factor describing the hydraulic behavior of a wetland. As D increases, the concentration of a certain pollutant decreases and thus leads to lower removal efficiency. Similarly, the Peclet number, a complementary measure of D defined as $P_e = 1/D = ul/d$ suggests that a large P_e value implies plug-flow conditions. That is, P_e values >500 indicate little dispersion from 40 to 500 moderate dispersion and values <40 suggest a large amount of dispersion and presence of mixed flow (Chazarenc et al. 2003). The value of t_{peak} is related to dispersion in the sense that an RTD function with a small peak time generally contains low dispersion. Also, the HLR may influence the dispersion in a CW, and studies have shown that the aspect ratio and depth are very important parameters for the dispersion processes (Persson and Wittgren 2003).

As discussed above, the CW hydraulic behavior is assessed based on the information extracted from a measured RTD curve. Spatial monitoring of tracer concentrations throughout the wetland identifies preferential flow paths and dead zones within CW (Grismer et al. 2003). The tracer is likely to be detected in preferential flow paths before it arrives in low-flow areas or dead zones within the wetland. The tracer is also likely to remain in backwaters and low-flow zones for longer periods when compared to preferential flow channels. Spatial tracer monitoring can be conducted along longitudinal, lateral, and vertical profiles within a wetland to gain an insight into the distribution of flow velocities, preferential flow paths, and mixing characteristics as wastewater moves through the system (Headley and Kadlec 2007; Galvão et al. 2010). A number of studies have used this technique in SSF wetlands to identify preferential flow across the bottom and therefore below the root zone (Su et al. 2009). Such studies can be used to gain an indication of the influence of inlet and outlet configurations (such as the vertical location of distribution pipes) on wetland flow paths. Headley et al. (2005) injected a tracer impulse at the mid-depth of an SSF-CW and monitored its progress at different depths downstream. They found the occurrence of substantial vertical mixing as water progressed through the system. There is a great opportunity for improving our understanding of flow dynamics and optimizing treatment wetland design through the execution of carefully conducted tracer studies.

SUMMARY

An optimal design model for constructed wetlands is expected to describe and estimate wetland hydraulics well since this directly influences treatment efficiency. A summary of the important fundamental elements of ideal plug flow theory is presented as a starting point to learn about hydrodynamics and mixing in chemical reactors.

Many of the important biogeochemical and chemical reactions during the wastewater passage depend on contact time between wastewater constituents and microorganisms and/or substrate. The main method, which gains information about internal hydraulic processes, is with inert tracers, which provides the capability to interpret hydraulic characteristics, such as residence time, degree of mixing, and short-circuiting. Any short-circuiting or dead zones that occur within a wetland will consequently have an effect on contact time and flow velocities, and therefore an impact on hydraulic/treatment efficiency. Hence, this chapter reviews and summarizes the hydraulic theory used when conducting hydraulic tracer tests in treatment wetlands and elaborates further on wetland hydraulic assessment, with a view to providing enough information for those without a background in chemical engineering.

CHAPTER 3

Hydraulic Design of Constructed Wetland

Saeid Eslamian, Saeid Okhravi,
and Mark E. Grismer

3.1 GENERAL

Constructed wetland (CW) design requires determination of
the necessary detention time for the desired treatment, associated CW volume/area, loading rates, substrate medium, and
water depth. The general CW design process typically is as
follows:

1. Set design requirements

    a. Determine design flow rates and associated CW area/
    volume

    b. Characterize influent wastewater and desired level of
    treatment

    c. Identify CW effluent discharge location and associated
    permit requirements

2. Obtain relevant hydrologic data and landscape information

   a. Historic daily precipitation, mean temperature, and evaporation rate data

   b. Soil types, topography, and prevailing native wetland plant species

   c. CW discharge conditions and constraints

3. Design pretreatment system for desired wastewater treatment

   a. Suspended solids removal (e.g., rotating screens, septic systems)

   b. Reduction of toxin and some nutrient concentrations as needed for the particular CW type planned and desired treatment level

   c. Wastewater pH buffering or supplemental micronutrients necessary for desired microbial treatment

4. CW design

   a. Select CW surface area and layout as determined by anticipated loading rates and available land area

   b. Determine CW configuration, including settling zone (if limited pretreatment) and aspect ratio depending on (a)

   c. Identify type of substrate media, gradation, hydraulic conductivity, and approximate CW water depth

   d. Consider and incorporate desired open water/vegetation ratio (FWS CWs)

   e. Determine required HRT for the desired level of treatment (including design safety factor) and prevailing water temperatures (seasonal HRT values)

f. Based on the required HRT, estimate allowable ranges of other hydraulic parameters and how they may be affected by various design options considered

g. Design inlet and outlet structures to meet the required hydraulic parameters

As discussed in the previous chapter, CW treatment performance depends on a relative uniform distribution of flow velocity across the CW substrate and/or surface. We consider the hydrodynamic design aspects of CWs relevant to obtaining uniform or near plug-flow conditions below.

3.2 BASIN DIMENSIONS

Generally, FWS CWs are much larger than SSF CWs; they have lesser treatment capacity and are subject to greater disturbances associated with the local climate conditions (Kadlec 2009). The wetland volume consists of the length-to-width ratio (L:W) and basin depth. CW size can be determined using Equation 3.1 (Kickuth 1984).

$$A_s = \frac{Q_d(\ln C_i - \ln C_e)}{K_{BOD}} \tag{3.1}$$

A_s is the surface area of bed (m^2); Q_d is the average daily flow rate of wastewater (m^3/d); C_i is the influent BOD concentration (mg/L); C_e is the effluent BOD concentration (mg/L); K_{BOD} is the rate constant (m/d); K_{BOD} is determined based on the expression $K_T h \phi$, where K_T is the K_{20} $(1.06)^{(T-20)}$; K_{20} is the rate constant at 20°C (d^{-1}); T is the operational temperature of the system (°C); h is the depth of water column (m); ϕ is the porosity of the substrate medium (percentage expressed as fraction).

K_{BOD} is temperature dependent, and the BOD degradation rate generally increases about 10% per °C and relative "maturity" (operational age) of the system (UN-HABITAT 2008). As a starting point in design lacking other degradation rate information, K_{20} is assumed to be equal to 1.1.

In addition to the required surface area for HSSF and VSSF, there is a unique bed cross-sectional area determined from Equation 3.2 (i.e., the Darcy equation for flow in porous media).

$$A_c = \frac{Q_a}{K(dH/ds)} \qquad (3.2)$$

A_c is the cross-sectional area of the bed (m²); Q_a is the average flow (m³/s), identical to Q_d; K is the hydraulic conductivity of the fully developed bed (m/s); dH/d_S is the bed bottom slope or grade (m/m).

In HSSF CWs, the CW depth can be determined by simultaneous solution of Equations 3.1 and 3.2. First, we calculate the required bed cross-sectional area (Equation 3.2) and then determine the CW width using the required CW substrate depth. CW length is calculated by dividing the plan area by the design CW width. Often, this involves a trial-and-error solution approach as some parameters have only preliminary values, or are unknown, and available CW land area may be a factor.

In general, the SSF CW substrate depth is approximately the plant rooting depth so that the roots are in contact with the flowing water and play a role in treatment or aeration. The typical operational water depth in FWS CWs is about 30 cm, at a water fraction ~95% on an area basis. For HSSF CWs, typical substrate depths are 0.5–1 m with a plant coverage fraction of ~60%. Assuming simple decay applies with a degradation coefficient of 1.1, the corresponding detention times are ~9 and 3 days for FWS and HSSF CWs, respectively. García et al. (2004) suggested that HSSF wetlands with an average depth of 27 cm might be more effective than deeper HSSF wetlands with an average water depth of 50 cm, but this depth leaves little allowance for changing climate (rain and evaporation), unanticipated flow variations, pore clogging, particulate/detritus settling, and plant rooting depth (Vymazal 2009).

VSSF CWs are generally built to greater depths as compared to HSSF systems because of their limited surface area. Most VSSF systems in UK have depths of 50–80 cm (Cooper et al. 1996). In contrast, Austria and Germany use depths of 0.8–1.0 m (UN-HABITAT 2008). Denmark requires a minimum depth of 1.0 m (Brix and Arias 2005), a depth also originally used in Nepal, though smaller depths are being used presently.

In subtropical climates, it is possible to increase the applied loading rates above those guidelines issued in Central Europe and achieve adequate nitrification in the VSSF CW (Philippi et al. 2006). VSSF CWs with a 0.75 m depth outperformed shallower beds of 0.45 m depth. The UN recommends a substrate depth of 0.70 m to provide sufficient nitrification in addition to the organic pollutant removal (UN-HABITAT 2008).

The CW aspect ratio is also an important design consideration, as aspect ratios between 3:1 and 5:1 are typically effective (EPA 2000). If such CW lengths are not possible due to site constraints, a minimum ratio of 1.9:1 should be considered together with the possibility of designing the CW along a curvilinear contour if possible (Su et al. 2009). Use of large aspect ratios greater than 10:1 required additional hydraulic headloss considerations in the design and possible flow stagnation in HSSF CWs.

3.3 SUBSTRATE

CW substrate is critical toward pollutant removal and adequate plant rooting conditions and directly affects the wetland hydraulics, particularly in the SSF systems. Depending on local availability and costs, substrate materials range from relatively inert sands, gravels, and drain rock or aggregate to more reactive (higher CEC[1] or specific surface) materials with internal pore space such as

---

[1] Cation Exchange Capacity.

zeolite aggregate (Collison and Grismer 2018a, 2018b). Materials with additional adsorption or exchange capacity or simply large specific surface areas can enhance the treatment capacity of the CW (Ren et al. 2007). The CW substrate media perform several functions:

- Provide environment for vegetation rooting

- Improve flow distribution through wetland cells

- Provide surface area for microbial growth and biofilms

- Filter particles and slow flow velocities to enable sediment settling.

Substrate hydraulic conductivity is an important design parameter that is largely controlled by sand/gravel/aggregate median grain size and degree of compaction or bulk density. Finer-textured substrates such as fine sand improve suspended solids (TSS) filtering/trapping, and general wastewater treatment but have relatively small hydraulic conductivities and are subject to pore clogging. Such CWs require smaller loading rates and must be designed with back-flushing capability to manage pore clogging and undesirable surface flows in SSF CWs. Larger size aggregate materials have much greater hydraulic conductivity and are less subject to pore clogging from excessive TSS loading (Naipal et al. 2014). On the other hand, very large particles have high conductivity but lower surface area per unit volume to support microbial biofilms limiting treatment potential. The compromise is for intermediate-sized materials (5–20 mm) generally characterized as gravels. Although a range of different grain sizes are recommended as substrate for CWs, some research suggests that fine gravels are more effective in pollutant removal than coarse gravels (Yousefi and Mohseni-Bandpei 2010). The choice of substrate gravels/aggregates is largely determined by local availability as transport costs are likely greater than that of the substrate.

However chosen, the gravel/aggregate should be thoroughly washed prior to placement in the CW basin so as to avoid later clogging problem and creation of low-flow or "dead zones" within the CW (Grismer et al. 2003).

3.3.1 Horizontal Subsurface Flow CW: Substrate Materials

In HSSF CWs, a wide range of particle sizes (<1–60 mm) have been used as substrate materials (Vymazal 1996; EPA 2000; UN-HABITAT 2008) and, with respect to treatment, there has been no clear advantage when using media in the 10–60 mm range (EPA 2000). Okhravi et al. (2017) experimented with 40 media samples in terms of best grain gradation and recommended fine grains between 3 and 12 mm with a coefficient of curvature $(C_c = d_{30}^2 / (d_{10}d_{60}))$ close to one as a substrate material for HSSF wetlands. The grain size selection for the CW system inlet where fluid dynamics can be initially turbulent with high dispersion could be different from the remainder wetland; however, careful design of the inlet manifold to evenly disperse flows across the width/depth of the CW usually manages this issue. To avoid clogging at the inlet and outlet zones, it is recommended that the media be 40–80 mm in diameter (coarse gravel).

3.3.2 Vertical Subsurface Flow CW: Substrate Materials

Selection of a suitably permeable substrate in relation to the hydraulic and organic loading is the most critical design parameter of VSSF CWs. Most treatment problems occur when there is insufficient permeability for the applied load, resulting in system ponding and water backup. Structurally, there should be a 20 cm deep layer of gravel/aggregate (a minimum of 20 cm) at the top and base of the system for drainage, with a 40–80 cm thick layer of finer sands and/or gravel (hydraulic conductivity, $K = 10^{-4}–10^{-3}$ m/s) in between for wastewater treatment. The top gravel layer contributes little to the initial filtering

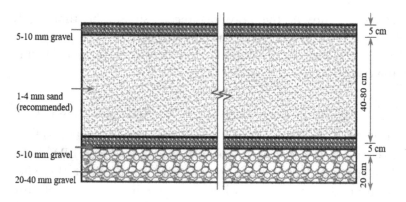

FIGURE 3.1 Recommended substrate arrangement in a VSSF CW. (From UN-HABITAT, *Constructed Wetlands Manual*, UN-HABITAT Water for Asian Cities Programme, Kathmandu, Nepal, 2008.)

process but promotes more uniform flow to the finer materials below and some additional storage capacity for fluctuating inlet flows (see Figure 3.1).

The substrate properties, d_{10} (effective grain size) and d_{60} and the uniformity coefficient (the quotient between d_{60} and d_{10}, $C_u = d_{60}/d_{10}$) are also important characteristics to consider when possible. The substrate should not contain loam, silt, or other fine material, nor should it consist of a material with sharp edges. Some literature suggests that for the treatment layer, the effective grain size should be $0.2 < d_{10} < 1.2$ mm with a uniformity coefficient $3 < C_u < 6$ (Reed 1990; Korkusuz 2005). Hoffmann et al. (2011) stated grain sizes with $0.1 < d_{10} < 0.4$ mm work well and recommended the use of sand (0–4 mm) as the main substrate with $0.3 < d_{10} < 0.4$ mm and $C_u < 4$.

3.4 BED/WATER SLOPE

Theoretically, the CW basin slope matches the water-level slope to maintain a uniform water depth across the bed. A practical approach is to uniformly grade the basin bottom along the

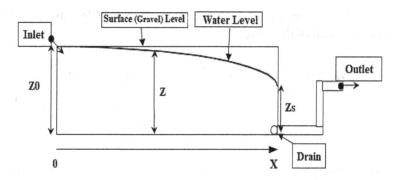

FIGURE 3.2 Water level within the bed in saturated horizontal flow. (Modified from Chazarenc, F. et al., *Ecol. Eng.*, 21, 165–173, 2003.)

flow direction from the inlet to outlet to facilitate draining when maintenance is required (Figure 3.2); basin bed slopes of 0.5%–2% are recommended, though there is little research directed at determining optimal grades (UN-HABITAT 2008). Similarly, the top surface of the media should be leveled or consistent with the basin slope for easier planting and routine maintenance. In FWS CWs, when the flow depth in the wetland increases, flow inundates shallow plants, increasing the flow velocities over those prior to inundations and recirculating flows into dead volume areas. However, as the flow depth increases further, the preferential flow paths tend to develop (short-circuiting), with a resulting decrease in the CW hydraulic efficiency.

The bed bottom slope is an important design consideration as it typically defines the gravitational gradient of the Darcy equation quantifying possible flow rates through the HSSF CW substrate. Hydraulic loading rates are limited to this design value to maintain subsurface flow conditions and limit surface flow short-circuiting. Writing the Darcy flux per unit cross-sectional area as $q = VA$ and $V = -K(dz/dx)$, where Q is the inflow in m^3 per day,

V is the flow velocity in ms^{-1}, and $A = wz$ in m^2 (flow area) and w is the width of the system:

$$Q = -K\frac{dz}{dx}wz \quad \text{and} \quad Q\int_0^x dx = -Kw\int_{z_0}^z zdz \quad \Rightarrow$$

$$z = \left[z_0^2 - \frac{2Q}{Kw}x \right]^{0.5}$$

(3.3)

Using Equation 3.3 to determine hydraulic retention time, we assume the inflow and filter porosity are constants and solve for the time below.

$$t = \frac{1}{Q}\int_0^x \phi wz dx = \frac{\phi}{Q}\int_0^x w\left(z_0^2 - \frac{2Q}{Kw}x \right)^{0.5}$$

$$= \frac{\phi w}{Q}\left[\frac{-Kw}{3Q}\left[z_0^2 - \frac{2Q}{Kw}x \right]^{1.5} \right]_0^x$$

(3.4)

$$t = \frac{\phi Kw^2}{3Q^2}z_0^3\left(1 - \left(1 - \frac{2QX}{Kwz_0^2} \right)^{1.5} \right)$$

Verification: if $\frac{2QX}{Kwz_0^2} \ll 1$ then, $t \approx \frac{\phi Kw^2}{3Q^2}z_0^3$, where K is the hydraulic conductivity in ms^{-1}; Q is the inflow and outflow rate in m^3s^{-1}; z_0 is the water level at the head of the filter in m; X is the length in m; and ϕ is the porosity. For example, Chazarenc et al. (2003) injected an impulse tracer during a hydraulic overload period with an inflow of 4.66 m^3h^{-1} (Q = 1.29 × 10^{-3} m^3s^{-1}) and computed a hydraulic retention time (t) between 27.1 and 34 hours. This model for experimental HRT determination supposes that infiltration, as well as hydraulic conductivity, is constant (homogeneous and isotropic) in the substrate across the CW. Assuming that all flows are maintained within the substrate, the required bed slope can be calculated from the Darcy equation in Equations 3.3 and 3.4.

3.5 BED SEALING

Ideally, CW cells are underlain with an impermeable liner material (e.g., compacted clays with $K < 10^{-7}$ m/s, plastic, pond geotextiles) whenever possible to limit possible inflow from adjacent areas or possible seepage of partially treated wastewater to the local groundwater system. Once operational and sometime after the porous media "ripens" and treatment improves, wetland liner or base soils tend to "seal" from accumulated detritus, particulate, and microbial exudates. European Guidelines (Cooper 1990) suggest that if the local soil has a hydraulic conductivity, $K < 10^{-8}$ m/s, then it is likely to have high clay content and could provide adequate bed sealing without an artificially placed liner. Otherwise, the following guidelines about additional liner materials based on the placement soil K values should be considered in the CW design (UN-HABITAT 2008):

- $K > 10^{-6}$ m/s $\Rightarrow$ Lining required for CW;

- $10^{-6} > K > 10^{-7}$ m/s $\Rightarrow$ Limited seepage may occur but future CW sealing likely;

- $K < 10^{-8}$ m/s $\Rightarrow$ Limited seepage and CW will seal naturally, no liner required.

When plastic, rubber, or other synthetic liner materials are available and required, the liners should cover the CW base and side walls; all joints and connections need to be sealed. Polyethylene liners must be UV resistant and individual liner sheets welded together. Plastic/vinyl liner thickness for use in CWs range from 1 to 3 mm with at least 2 mm thick liners recommended to minimize placement leakage from small punctures or tears. Thicker liners are not recommended as they are more difficult to place and seal. The native soil subgrade below the liner should be finished as smoothly as possible and on the design slope (0.5%–3%) prior to liner placement. Liner materials should be laid out as smooth as possible in warm weather with no additional tension to allow flexibility when filling with substrate aggregate. Liner should extend over the side berms of the CW and tested for leakage after placement and sealing by partial filling of the basin with water prior to addition of aggregate.

3.6 INLET–OUTLET ARRANGEMENT

As Suliman et al. (2006) noted, inlet and outlet control structures have a great influence on the CW hydraulic behavior. Inlet structures distribute the flow onto and across the wetland width and influence the eventual flow path through the wetland, while outlet structures control CW water depth and width of collection. Multiple inlets and outlets spaced across either end of the wetland are essential for creating uniform influent distribution into and flow through the wetland. These structures help to prevent "dead zones" where water exchange is poor, resulting in effective wastewater retention times less than design values and poorer treatment than expected. The inlet/outlet structures should minimize the potential for short-circuiting and clogging in the media and maximize even flow distribution, while the outlet structure should maximize even flow collection and allow water-level control within the CW and bed drainage as needed.

The inlet control structures for CWs include surface and subsurface manifolds such as perforated pipes, or open trenches oriented perpendicular to the CW flow direction that uniformly distribute the inflow across the entire CW width. Su et al. (2009) and Okhravi et al. (2017) investigated the effects of different inlet layouts on the internal hydraulic behavior of FWS and HSSF CWs. Both studies clarified that uniformly distributed inflow would be the best arrangement by all performance criteria. Manifold pipe dimensions, orifice diameters and spacing depend on the design inflow rate. Where possible, the inlet manifold is installed near the aggregate surface to allow access by the operator for flow adjustment and maintenance. A subsurface manifold avoids build-up of algal slimes and the consequent clogging that can occur next to surface manifolds, but are more difficult to adjust and maintain.

Uniform distribution of flows is also critical to VSSF CWs where the flow is distributed intermittently from up to down by the network of pipes with downward pointing holes. The pipe ends should be raised so that air can pass through during flushing as well as to achieve equal distribution of the wastewater.

Similarly, CW outlet control structures should uniformly collect wastewater effluent across the entire width of the wetland. The SSF design should allow controlled flooding to depths of 0.1–0.2 m to foster desirable plant growth and control weeds. Installation of valves or gates at the CW outlet that enable control of CW water levels to ensure both adequate hydraulic gradient across the bed and have significant benefits in operating and maintaining the wetland. In HSSF CW systems, the outlet is often a perforated drain pipe enclosed in a 0.5 m wide drainage zone filled with large graded stones. This leads to a sump where the water level is controlled by either a swiveling elbow or a socketed pipe (UN-HABITAT 2008). In VSSF CW systems, the collection system may consist of a network of drainage pipes surrounded by large stones. The drainage pipe leads to a collection sump that allow the vertical bed to completely drain when needed.

3.7 VEGETATION

Properly designed CWs incorporate a vegetation zone where influent wastewater is introduced to the wetland. This is to promote emergent vegetation growth and subsequently increase solids flocculation and separation. Vegetation and related litter are necessary for the successful performance of CWs and contribute aesthetically to their appearance. Selection considerations for the CW vegetation to be planted include

- Use of locally dominant macrophyte species
- Species with deep root penetration, strong fibrous root and rhizomes
- Species with high stem densities

- Species with high root densities to provide surface area for biofilms

- Species with oxygen transport capacity into the root zone to facilitate oxidation of reduced toxic metals and support a large rhizosphere.

Several commonly used CW plant species include *P. australis* (Common Reed), *Phragmites* sp., and *Typha* sp. due to their wide climate tolerance and rapid growth.

Establishing vegetation within a CW involves placing suitable plants in the spring season, most likely leading to their successful establishment prior to addition of wastewater flows to the CW. Commonly used planting densities range from 0.3 to 1.0 m on center depending on costs, availability of plant stock and time required before introduction of wastewater flows to the CW. Of course, the greater the planting density, the more rapid the system development, but at greater initial construction cost. Only a small fraction of the ultimate plant density is required, again, depending on the rate of plant reproduction and the acceptable timeframe for plant establishment. Successful incorporation of plants within the CW requires the following (Wallace and Knight 2006):

- Plant species matched to the wetland water/flow regime

- Sufficient viable plant material and available planters at the time of the planting

- Water level management during startup compatible with the plants establishment needs.

For FWS CWs, the water level should just cover the top of the rooting material. The plants need access to air, and if new plants are inundated for extended periods, there will be inadequate stand establishment in the CW. Water depths can be increased as plants grow. For SSF CWs, the water level should be just over the gravel bed, but not above the vegetation with the goal of providing a continuous water supply to the plant root zone during establishment.

Routine maintenance of wetland vegetation is not required for CW systems operating within their design parameters. Wetland plant communities are self-maintaining and will grow, die, and regrow every year. If plants naturally spread to areas not intended for vegetation, the plants should be harvested (EPA 2000). Plant harvesting can affect CW hydraulic performance, so the harvested cell should be removed from service before and after harvesting. Harvested CW vegetation can be burned, chopped, and composted, used as mulch, or digested. Vegetation and/or debris may affect substrate maintenance of CWs through clogging that may require additional measures including shallow ripping of the substrate to restore hydraulic performance.

SUMMARY

This chapter has been prepared as a general guide to the design of constructed wetland with a great emphasis on hydraulic characteristics. The general step design process has been presented in the first section of the chapter. The following sections explain the design considerations required for six major hydraulic design components (basin, substrate, bed/water slope, bed sealing, inlet–outlet arrangements, and vegetation) of this engineered system in detail.

CHAPTER 4

Factors Affecting Constructed Wetland Hydraulic Performance

Saeid Okhravi, Saeid Eslamian, and Mark E. Grismer

4.1 GENERAL

As with any wastewater treatment system, a proper functioning and operable CW depends on many factors specific to the location, wastewater type, variable hydrologic conditions, among other factors. Herein, we consider some of the more common issues found to impact CW hydraulic performance as they directly influence CW treatment efficiency. Design parameters such as aspect ratio, size of the porous media, water level, inlet–outlet configurations, and hydraulic loading rate can improve the CW hydraulic performance such that ideal, "plug" or uniform flows occur across the system. Recall that plug flow implies minimal dispersion, dead volume areas, and short-circuiting that all adversely affect wastewater treatment through the reduction of design residence times.

As described in the tracer study Section (2.2.1), knowledge of the wastewater RTD within the CW enables description of how water moves through the wetland cells and determines its hydraulic and treatment performance. Actual residence times within a CW are often less than the design values likely due to establishment of preferential flow paths resulting in "dead" areas where flow is limited, or less than the average bulk flow rate.

4.2 WATER BALANCE CONSIDERATIONS

Careful attention to SSF CW hydrology is critical for obtaining satisfactory wetland function. From a water balance perspective, the flow entering a wetland is rarely equal to that exiting the wetland. During dry periods, CW inflows will exceed outflows due to evapotranspiration (ET) losses resulting in greater outflow salinity, while during rainy periods, outflows may exceed inflows, thereby decreasing outflow salinity. Evapotranspiration is the combined water loss by evaporation from substrate and plant leaf surfaces as well as through plant transpiration. As ET rates are diurnal, peaking during the early afternoon and minimal during the nighttime hours, CW discharge declines to minimum values during mid-afternoon and increase to maximum values prior to sunrise if there is no rain (Wallace and Knight 2006). For FWS CWs, specific ET rates have been consistent with prevailing alfalfa reference ET (ET_o), or about 10% greater than grass ET_o, or simply assumed to be about 75% of local Class A pan evaporation (EPA 2000). As with wetland ET, precipitation must also be included within the overall CW water balance and used to determine appropriate hydraulic retention times. A properly designed CW has sufficient perimeter berms to limit runoff from the surrounding area from entering the CW, but direct precipitation into the CW can have a direct impact on treatment process (EPA 2000).

Without adequate consideration of the CW hydrologic (water) balance conditions in the design, the proposed wetland will invariably have operational and performance issues. Excess, unaccounted for water entering the system will limit settling and increase discharge rates, while too little water into a wetland will stress vegetation resulting in potential loss of vegetative treatment capability. Both conditions reduce wastewater treatment from design objectives.

As wetland flow characteristics are highly influenced by the shape and bathymetry of the CW, the hydraulic efficiency changes as the water level rises and internal embankments become flooded. The phenomena is of likely more concern in CW systems treating stormwater runoff and subject to widely variable inflows and hydrologic conditions as compared to wastewater treatment CWs receiving regular inflows. As the wetland water depth increases, flow begins to inundate the shallow marshes in the FWS systems, which initially redistributes flow into the deeper pools, increasing the hydraulic efficiency of the system. However, as the depth continues to increase, the flow tends to short-circuit, with a resulting decrease in the hydraulic efficiency (Jenkins and Greenway 2007). Generally, as CW water levels increase, the effective hydraulic loading rate also increases and may result in flow short-circuiting and decreased treatment performance.

4.3 ASPECT RATIO

The L:W aspect ratio is a key design factor that can help increase the hydraulic efficiency of the CW. As the aspect ratio increases, actual retention times typically approach theoretical plug flow values, resulting in more uniform flows and less dead volumes within the CW. On the other hand, increased aspect ratio leads to increased dispersion along the length of the CW and may have an effect on wastewater treatment in the relatively short term, but little effect on the overall average treatment performance after more than one pore volume has passed through the CW (Alcocer et al. 2012). Improvement of the CW hydraulic behavior with increases in aspect ratio is likely attributable to larger distance and travel time allowing initial flow turbulence to dissipate.

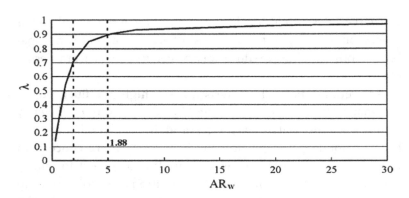

FIGURE 4.1 Relationship between the hydraulic efficiency and the aspect ratio of FWS wetlands. (From Su, T.M. et al., *Ecol. Eng.*, 35, 1200–1207, 2009.)

Figure 4.1 illustrates the relationship between the AR_w (aspect ratio) and λ (the ratio of t_p, given by RTD curve to nominal retention time, t_n) observed in a FWS CW. As mentioned previously, λ increases with the AR_w due to increased dispersion, resulting in higher t_p but the rate of λ increase greatly declines at $AR_w > 5$, suggesting limited practical effects of AR_w on λ for $AR_w > 5$. Therefore, if field conditions permit, CW design should consider using an $AR_w \sim 5$ because it likely will achieve a $\lambda \sim 0.9$ or higher. If this is not possible, then every effort should be made in the CW design to have $AR_w > 1.9$ so $\lambda > 0.7$ (Su et al. 2009).

4.4 HYDRAULIC LOADING RATE

Figure 4.2 illustrates how influent loading rate effects the CW retention time distribution; generally, increased hydraulic loading rates (HLRs) within the CW design capacity results in more uniform flows and actual retention times approach plug flow values as shown in this example, where the curve centroid shifts toward the normalized time of 1.0 (Alcocer et al. 2012). Increased HLRs also increase dispersion in a CW and may decrease treatment efficiency (Zahraeifard and Deng 2011).

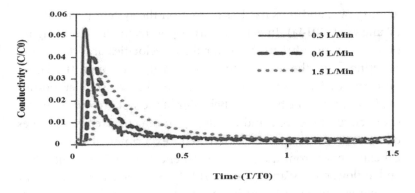

FIGURE 4.2   The RTD curve of the three different influent hydraulic loading rates with aspect ratio 1:1. (From Alcocer, D.J.R. et al., *Ecol. Eng.*, 40, 18–26, 2012.)

Excessive (beyond design) inflow rates can result in preferential flow also lowering treatment performance. This effect can be dampened by increasing the aspect ratio as the effects of turbulence and dispersion may be dissipated. Fundamentally, the CW design should allow for a range of HLRs such that the system behaves like a series of completely mixed reactors (CSTR) and achieves the desired treatment. HLRs that exceed the design capacity can be expected to result in decreased CW treatment performance.

4.5 SUBSTRATE SIZE

The substrate porous media size and its placement hydraulic conductivity controls the internal hydrodynamic behavior of CW. Flow through smaller diameter media generally decreases the transverse cross section available for fluid flow, thus increasing the linear velocity of the fluid. Consequently, the use of smaller substrate materials had the same effect as increasing the aspect ratio or the influent loading, leading to increased dispersion and broader RTD. The effect of substrate size on the plug flow and dead volume ratios is less pronounced than the effects of the aspect ratio and influent loading rate, and thus has less impact on the system performance (Wang et al. 2014a). In general, larger porous media yields higher plug flow ratios due to the lower linear velocities and dispersion resulting from the less constricted fluid flow through the larger pore spaces (Alcocer et al. 2012). Nevertheless, smaller porous media improves the hydraulic behavior of the system by the smaller short-circuiting effects and development of actual retention times closer to theoretical values. Based on the experimental and numerical experience from the combination effects of different filter size and inflow rate, Wang et al. (2014b) concluded that the inflow discharge rate has a larger impact on plug flow ratio than size of substrate material indicated at bar chart Figure 4.3. A decrease in inflow loading rate typically results in plug flow behavior. The fluid parcels of wastewater passes fast due to high inflow rate leading to preferential flow paths and lowering flow exploitation on active zones in porous media. For example, when the inflow increase from 5 to 15 mL/s, a decrease of plug flow value ranged from 98% to 93% of the total porous volume. Fine filter size has not enough porous space compared to coarser media, resulting in the reduction of plug flow ratio or effective volume at the same inflow rate.

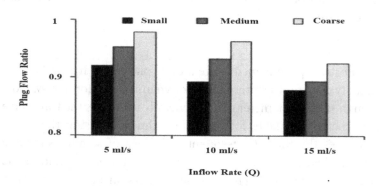

FIGURE 4.3 Plug flow ratio for different inflow rates and filter sizes. (Adapted from Wang, Y. et al., *Ecol. Eng.*, 69, 177–185, 2014b.)

4.6 INLET/OUTLET CONFIGURATION

CW hydraulic performance can be enhanced by appropriately managing flow distributions at the inlet and outlet. Using a 3D simulation combined with a tracer study, Okhravi et al. (2017) examined three inlet–outlet design configurations: midpoint–midpoint (A), corner–midpoint (B), and uniform–midpoint (C) in the same wetland. The effects of the three different inlet-outlet layouts on the CW RTD are shown in Figure 4.4. The tracer HRTs were 4.53, 3.24, and 4.65 days at midpoint–midpoint (A), corner–midpoint (B), and uniform–midpoint (C), respectively; this was much less than the nominal HRT of each configuration (5.17, 5.22, and 5.32 days). Most tracer studies result in a mean hydraulic retention time less than the nominal retention time due to the tracer loss and the inevitable tail in the RTD curve.

Generally, larger HRTs imply that the CW system has more time to treat the influent (Bruun et al. 2016). The results indicated that configuration C had the maximum retention time (4.65 days). The overall hydraulic results of this study are summarized in Table 4.1.

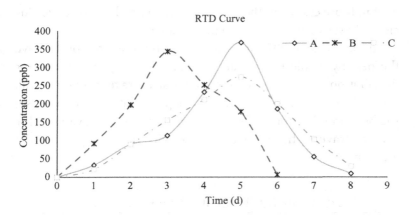

FIGURE 4.4 Measured RTD curve at three different cases of inlet, midpoint (A), corner (B), and uniform (C). (From Okhravi, S. et al., *Ecohydrol. Hydrobiol.*, 17, 264–273, 2017.)

TABLE 4.1 Derived hydraulic Parameters from the Different Cases on Inlet and Outlet Setup, Midpoint-Midpoint (A), Corner-Midpoint (B), and Uniform Midpoint (C)

% λ	e_v	S	t_m (d)	t_{50} (d)	t_{16} (d)	HLR (cm/d)	t_n (d)	Q (m³/d)	Case
96	0.875	0.58	4.53	4.23	2.45	6.33	5.176	6.58	A
57	0.621	0.51	3.24	2.7	1.4	6.27	5.224	6.52	B
94	0.875	0.57	4.65	4.2	2.41	6.15	5.322	6.4	C

Source: Okhravi, S., et al. *Ecohydrol. Hydrobiol.*, 17, 264–273, 2017.

In cases A and C, which had the least short-circuiting (i.e., high S values), there was greater flow uniformity and hydraulic efficiency, as compared to case B, which had the worst inlet configuration, largest dead-volume ratio, and lowest hydraulic efficiency. The plug flow fraction or effective volume in cases A and C were equal (~87.5%), indicating a dead volume of 12.5%, as compared to 38% in configuration B. Further, when hydraulic efficiency exceeded 0.7, the system had uniform hydraulic conveyance, thus reducing the risk of short-circuiting (see Section 2.3). Cases A and C have similar hydraulic efficiency. A little difference between flow rates at the inlet in those cases and their effects on λ could alter hydraulic efficiency in the case A more than in the case C.

Suliman et al. (2007) noticed that bottom inlet–top outlet have the best hydraulic performance by producing greater width for distribution of transition times, longest average retention time due to longer flow path, and lowest pore water velocity. Also, the bottom inlet–top outlet configuration has the advantage for flow to traverse the filter against gravity increasing hydraulic efficiency.

4.7 HYDRAULIC MODELING AIDED DESIGN

Hydraulic behavior of CW can be evaluated not only through tracer tests (Headley and Kadlec 2007; Kadlec 2007) but also by numerical–mathematical modeling (Martinez and Wise 2003;

Grismer 2005; Marsili-Libelli and Checchi 2005; Rajabzadeh et al. 2015). Because of the great variety of CW typologies, applications, and processes involved in the treatment, numerical models were developed and refined in order to assist in CW design and performance evaluation. However, most design models deploy typical parameters derived from pilot-scale CWs experiments. Therefore, the model may be valid only for boundary conditions under which the experiments were carried out like flow discharge, hydrologic conditions, influent composition, substrate material, plant species, inlet–outlet setup, etc. (Rousseau et al. 2004). Strong efforts are underway to develop or apply compartmental mechanistic models to explain and predict CW behavior, especially internal hydraulic processes.

A literature review of existing simulation tools for CWs indicated that observed CW hydraulic behavior can be readily matched with simulated data through modification of uncertain effective parameters (Langergraber 2011). Using Hydrus2D simulation, Langergraber (2008) reproduced the hydraulic head distribution along the CW and concluded that the hydraulic properties as compared to biokinetic parameters have more impact on model calibration. Further, he found that a good match between experimental data and predicted CW treatment performance could be achieved by first developing a well-calibrated hydraulic model and then using default reactive transport parameters for the biokinetic model.

Galvão et al. (2010) developed a one-dimensional dynamic model for hydraulic design in HSSF CWs. The hydraulic module assumes Darcy's law for computing head losses along the substrate porous media and the CW outlet boundary condition. Further, the model includes the hydrologic parameters, precipitation, temperature, and evapotranspiration disregarded in previous developed models. Measured data from a CW operating in the South of Portugal was used to calibrate and successfully validate the model. They found that evapotranspiration patterns were a critical part of model validation success.

Other researchers have focused on development of hydraulic models to characterize CW flow regimes. Fioreze and Mancuso (2019) combined ModFlow and ModPath to simulate flow patterns through a HSSF CW with changes in the hydraulic conductivity of the porous medium and inflow distribution. The model proved to be a powerful tool for 3D simulation, allowing the representation of flow distribution, flow velocities, hydraulic head, and particle trajectories. They found that coarse sand yielded higher plug flow ratios and that the inlet setup can contribute to more uniform flows in the porous media. Samsó and García (2013), Okhravi et al. (2017), and Rajabzadeh et al. (2015) used a robust computational fluid dynamics (CFD) 3D model built on the Comsol Multiphysics platform to show internal hydraulic patterns effective in treatment processes in HSSF and VSSF CWs. The model depicted low-velocity zones and readily simulated measured hydraulic data, since the model was calibrated against experimental tracer data that had been collected in a full water-recirculation batch operation for all studies. This model also integrates hydrodynamics phenomena with biokinetics affecting contaminant removal and biofilm development and detachment similar to realistic field conditions. The main difference in model formulation between Samsó and García (2013) and Okhravi et al. (2017) with Rajabzadeh et al. (2015) was the governing equation used to adjust velocity and pressure profiles of flow in porous media. The first two employed the simpler Darcy equation, a linear relationship between hydraulic gradient and flow velocity in porous media, while Rajabzadeh et al. (2015) used the Navier–Stokes equations for describing the flow in freewater surface regions, and the Brinkman equations for the flow through the porous substrate. The ignored effects of viscous and inertia terms in Darcy's law are included in the momentum balance model of Brinkman. Kadaverugu (2016) used an open-source CFD environment-OpenFoam to simulate dynamic behavior of HSSF CW considering the nonlinear hydrodynamics and contaminant transport. Their model coupled variably saturated flow in the substrate material with the advection, dispersion, and decay of biological pollutants. This model was able to visualize pollutant transport in space and time; a helpful process for environmental designers to better design CW treatment systems.

Such numerical modeling will be valuable in improving our understanding of flow–substrate–vegetation–pollutant interactions and should enable better description of CW hydraulics and anticipated treatment performance that ultimately results in more cost-effective designs.

4.8 DESIGN FINDINGS SUMMARY

Design parameters such as CW aspect ratio, hydraulic loading rate, substrate material and dimensions, operational water level, and inlet–outlet configurations have direct impacts on CW hydraulic performance. Herein, we summarized the effects of these design factors to recommend the best model integration of design parameters. Considering the "plug" or uniform flows concept (see Figure 4.5), with minimal dispersion, dead volume areas, and short-circuiting effects, it can be concluded, that in the CWs design, higher aspect ratios yielded more ideal hydraulic behavior with smaller dead zones and less preferential flows. Higher inflow loading rates increase dispersion and decrease plug flow than lower loading rates. Nevertheless, higher influent loading rates yield actual residence times, t_m, that are

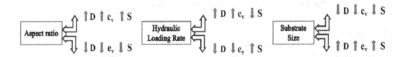

Knowing plug flow with less pronounced dispersion ($\downarrow$ D), dead volume areas ($\uparrow$ e_v) and short-circuiting effects ($\uparrow$ S)

Note: high S value (i.e., small short-circuiting effects) and high e_v value (i.e., lower dead volume areas)

FIGURE 4.5 Illustrating the effect of different design variables on plug flow condition.

closer to the nominal residence time, t_n. In regards to size of porous media, although larger porous media obtained greater plug-flow ratios and lower dispersion, smaller media decreases short-circuiting effects, and yields t_m closer to the t_n. In addition, not only did finer-textured media improve the surface area for biological growth, but they also did enhance the filtering/trapping potential of suspended solids (TSS) (Alcocer et al. 2012).

In HSSF CW design, the use of higher loading rates is constrained by the boundary limitation between the laminar and turbulent flow. In this case, the use of finer porous media dissipates the effects of turbulence and dispersion governing wetland behavior as a series of completely mixed reactors. Meanwhile, this acts like an optimum point that would increase surface area for vegetation growth, resulting in greater residence times and flow distribution, while minimizing pore-clogging risk associated with filtering and biological growth (Lee and Shih 2004). It should be noted that the plug-flow effects (dead volume, short-circuiting, and dispersion) would be augmented as scale effects of the experiments; hence, the effects of dead volume, short-circuiting, and dispersion observed in pilot-scale studies would be dampened in an anticipated larger operational systems. For this reason, Ioannidou and Pearson (2018) pointed out that hydraulic optimization guides do not necessarily conform to hydraulic parameters and values obtained from larger full-scale CWs, indicating the need to investigate more full-sized CW sites (e.g., Grismer et al. 2003).

SUMMARY

Previous works of literatures stated that the importance of the hydraulic design of constructed wetland cannot be overemphasized, and that short-circuiting is one of the greatest hindrances having successful constructed wetland design. Design parameters such as water level, aspect ratio, hydraulic loading rate, substrate size, and inlet/outlet configurations can improve the hydraulic behavior of constructed wetland systems by imparting an optimal plug flow regime through constructed wetland residing actual residence time close to nominal residence time. Hence, this chapter elaborates to assess the influence of design parameters on the plug flow and dead volume ratios to provide a further understanding of the internal flow pattern of constructed wetland. Also, a state of the art of numerical simulations' role for acquiring internal hydraulic information is presented.

CHAPTER 5

Future Constructed Wetland Research Orientations

Saeid Eslamian, Saeid Okhravi, and Mark E. Grismer

5.1 GENERAL

Constructed wetlands have been widely used across the world for a range of water treatment applications. CWs have great potential for decentralized wastewater treatment in small towns and rural areas with low polluted water or lake pollution control because of their relatively low costs for construction, operation, and maintenance. However, influent nature and the various contaminants will determine the optimal type of CW chosen and whether technical guidance is needed (Li et al. 2018). Selecting an appropriate CW requires knowledge of residence-time distribution and pollutant removal processes controlled by the CW hydraulics. Thus, this book specifically aims to assess CW hydraulic design aspects and integrated relations to obtain the best CW hydraulic performance and improve CW treatment efficiency.

The treatment achievable by a CW system can be optimized by addressing factors like the CW geometry, vegetation, the hydrologic and hydraulic effectiveness. In accordance with the study by Zhi and Ji (2012), between 1991 and 2011 articles about CWs have been published in at least 80 countries, with Europe, North America, East Asia, and Oceania being the four dominant regions producing these publications. Numerous studies have been published to cover various aspects of CWs, including insights on hydrological, physical, and biochemical processes. With growing number of scientific papers, the review papers and books try to summarize the design, development, and performance of different types of CWs. However, to the best of our knowledge, few attempts have been made to review hydraulic design aspects of CWs. Consequently, current hydraulic research emphases remains unclear. In this chapter, we suggest some of the potential current or future research should be directed at the hydraulic behavior effects of the CWs on the treatment performance for a variety of applications.

5.2 QUANTIFYING VEGETATION IMPACTS ON HYDRO-ENVIRONMENT MODEL DESIGN

Analysis of CW hydraulic performance is often performed by studying water flow patterns or HRT distributions (RTD) that describe the time duration that water parcels remain in wetland cells. An RTD curve reflects the degree of fluid mixing within a basin indicating short-circuiting, dead volume, dispersion, and other flow dynamics. Significantly, the distribution of residence times may depend on distributed differences of flow velocities in vertical or horizontal directions. Comprehending solute mixing within vegetation including nutrient transport is also vital in CWs affecting its hydraulic assessment (Lucke et al. 2019). Quantification of vegetation effects on hydro-environmental aspects of CWS remains quite challenging. Vegetation is not like the solid rigid unit on the system and characterized by plant root, leaves, and stems, partially saturated in water bodies, resulting in a multiscale problem that cannot be readily modeled. For this case, an approach that presents the average impacts of vegetation, such as a bulk mixing characterization, is generally preferred (Xavier et al. 2018). From a hydraulics perspective, CFD tools are a potential solution to improve CW hydraulic designs. Computational modeling can be used to simulate the flow through porous media integrating root zones with different stem densities and to analyze the impact of the different vegetation configurations on pollutant removal within individual root zones. An integrated approach that combines modeling with an instrumented field installation would be effective to enrich knowledge of the vegetation physical impact in internal hydraulic behavior of CWs.

5.3 INTERNAL HYDRAULIC FLOW PATTERN DETECTION

Water flow patterns through CWs are complex because of the large areas of plants and artificial matrix and influenced by several design parameters. A combination of these factors cause different hydraulic conditions inside the CWs, and the RTD provides useful insights about the CW hydrodynamics following construction and after operation. Measured RTDs within CWs are dynamic and difficult to predict *a priori*. Additional technologies are needed to study hydraulic flow patterns and hydraulic behavior of CWs in situ and models developed to predict the dynamic nature of RTDs associated with CW maturation. For example, Sun et al. (2016) proposed a technology based on the isotopic composition variation in CWs to detect the hydraulic flow patterns synchronously along the wetland. They determined the locations of preferential flow areas and dead zoned of CWs using stable hydrogen and oxygen isotopes. The affordable isotopes technology employs the internal elements in CWs to detect the hydraulic flow patterns in real time with no secondary pollution.

5.4 MINIMIZING SHORT-CIRCUITING EFFECTS BY INTERCONNECTED MECHANISMS

While the effects of different design variables such as aspect ratio, size of the porous media, and hydraulic loading rate on "short-circuiting" have been evaluated (see Figure 4.5), more work is needed. One finding of these studies was that an increase in aspect ratio yielded less short-circuiting and greater treatment performance. Similar results were obtained by decreasing HLR and substrate size separately. However, these predictions may not be the same for plug-flow behavior, since the plug-flow condition is characterized by the combination of less short-circuiting, dead volume, and dispersion. Designing a CW system to maximize actual residence time and prevent short-circuiting would improve the system effectiveness. It may also lead to a more efficient design by reducing the overall CW size. A number of innovative CW designs have focused on reducing short-circuiting by providing more contact time for the flow exposed to the vegetation zone and substrate to maximize treatment efficiency. For example, Walker et al. (2017) demonstrated the novel use of baffle/obstacle curtains to prevent short-circuiting of flows, which may be valuable to future studies in existing FWS and HSSF CWs.

The book found that hydraulic behavior of CWs is critical to effective design and the performance evaluation of CWs. Future designs would benefit by greater consideration of hydrodynamic design aspects in both computational modeling and full-scale CW applications.

SUMMARY

The rapid growth in published manuscripts addressing constructed wetlands indicates the promise of more papers and research in the near future. Understanding the hydraulic flow pattern within constructed wetland systems has received more attention during the last decade than it already has and therefore is predicted to be a primary research emphasis in the near future, as hydraulic efficiency is a key element in constructed wetlands technology. In this chapter, we suggest some of the current research and predicted future research directions in the development and enhancement of constructed wetland technology to improve the future hydraulic design of constructed wetlands.

References

Alcocer, D. J. R., Vallejos, G. G., & Champagne, P. (2012). Assessment of the plug flow and dead volume ratios in a sub-surface horizontal-flow packed-bed reactor as a representative model of a sub-surface horizontal constructed wetland. *Ecological Engineering*, 40, 18–26.

Ávila, C., Pelissari, C., Sezerino, P. H., Sgroi, M., Roccaro, P., & García, J. (2017). Enhancement of total nitrogen removal through effluent recirculation and fate of PPCPs in a hybrid constructed wetland system treating urban wastewater. *Science of the Total Environment*, 584, 414–425.

Brix, H. (1997). Do macrophytes play a role in constructed wetlands treatment? *Water Science and Technology*, 35 (5), 11–17.

Brix, H., & Arias, C. A. (2005). Danish guidelines for small-scale constructed wetland systems for onsite treatment of domestic sewage. *Water Science and Technology*, 51 (9), 1–9.

Bruun, J., Pugliese, L., Hoffmann, C. C., & Kjaergaard, C. (2016). Solute transport and nitrate removal in full-scale sub-surface flow constructed wetlands of various designs treating agricultural drainage water. *Ecological Engineering*, 97, 88–97.

Caselles-Osorio, A., Porta, A., Porras, M., & García, J. (2007). Effect of high organic loading rates of particulate and organic matter on the efficiency of shallow experimental horizontal sub-surface-flow constructed wetlands. *Water, Air, and Soil Pollution*, 183 (1–4), 367–375.

Chang, N. B., Xuan, Z., & Wanielista, M. P. (2012). A tracer study for assessing the interactions between hydraulic retention time and transport processes in a wetland system for nutrient removal. *Bioprocess and Biosystems Engineering*, 35 (3), 399–406.

Chazarenc, F., Merlin, G., & Gonthier, Y. (2003). Hydrodynamics of horizontal sub-surface flow constructed wetlands. *Ecological Engineering*, 21 (2–3), 165–173.

Clark, M. M. (1996). *Transport Modeling for Environmental Engineers and Scientists*. John Wiley & Sons, New York.

Collison, R., & Grismer, M. E. (2018a). Upscaling the Zeolite-Anammox process: Treatment of secondary effluent. *Water*, 10 (3), 236.

Collison, R., & Grismer, M. E. (2018b). Upscaling the Zeolite-Anammox process: Treatment of high-strength anaerobic digester filtrate. *Water*, 10 (11), 1553. doi:10.3390/w10111553.

Cooper, P. F. (1990). *European Design and Operations Guidelines for Reed Bed Treatment Systems*. Water Research Center, Swindon, UK.

Cooper, P.F., Job, G. D., & Green, M. B. (1996). *Reed Beds and Constructed Wetlands for Wastewater Treatment*. Water Research Center, Swindon, UK.

Crites, R. W. (1988). *Manual Design: Constructed Wetlands and Aquatic Plant Systems for Municipal Wastewater Treatment*. US Environmental Protection Agency, Office of Research and Development, Center for Environmental Research Information, Washington, DC.

Dierberg, F. E., & DeBusk, T. A. (2005). An evaluation of two tracers in surface-flow wetlands: Rhodamine-WT and lithium. *Wetlands*, 25 (1), 8–25.

EPA. (2000). Manual: Constructed Wetlands Treatment of Municipal Wastewaters. EPA/625/R-99/010, USEPA Office of Research and Development, Cincinnati, OH, October 2000, 154 pp.

Eslamian, S. (Ed.). (2016). *Urban Water Reuse Handbook*. CRC Press, Boca Raton, FL.

Eslamian, S., Okhravi, S., & Eslamian, F. (2014). Groundwater-surface water interactions. In *Handbook of Engineering Hydrology*, edited by S. Eslamian (pp. 259–287). CRC Press, Boca Raton, FL.

Eslamian, S., Okhravi, S., & Reyhani, M. N. (2016). Urban water reuse: Future policies and outlooks. In *Urban Water Reuse Handbook*, edited by S. Eslamian (pp. 1143–1151). CRC Press, Boca Raton, FL.

Fioreze, M., & Mancuso, M. A. (2019). MODFLOW and MODPATH for hydrodynamic simulation of porous media in horizontal sub-surface flow constructed wetlands: A tool for design criteria. *Ecological Engineering*, 130, 45–52.

Flores, L., García, J., Pena, R., & Garfí, M. (2019). Constructed wetlands for winery wastewater treatment: A comparative life cycle assessment. *Science of the Total Environment*, 659, 1567–1576.

Fogler, H. S. (1992). *Elements of Chemical Reaction Engineering*, 3rd ed. Prentice Hall, New Delhi, India.

Galvão, A. F., Matos, J. S., Ferreira, F. S., & Correia, F. N. (2010). Simulating flows in horizontal sub-surface flow constructed wetlands operating in Portugal. *Ecological Engineering*, 36 (4), 596–600.

García, J., Aguirre, P., Barragán, J., Mujeriego, R., Matamoros, V., & Bayona, J. M. (2005). Effect of key design parameters on the efficiency of horizontal sub-surface flow constructed wetlands. *Ecological Engineering*, 25 (4), 405–418.

García, J., Chiva, J., Aguirre, P., Álvarez, E., Sierra, J. P., & Mujeriego, R. (2004). Hydraulic behaviour of horizontal sub-surface flow constructed wetlands with different aspect ratio and granular medium size. *Ecological Engineering*, 23 (3), 177–187.

Grismer, M. E. (2005). Simulation evaluation of the effects of non-uniform flow and degradation parameter uncertainty on subsurface-flow constructed wetland performance. *Water Environment Research*, 77 (7), 3047–3053.

Grismer, M. E., Carr, M. A., & Shepherd, H. L. (2003). Evaluation of constructed wetland treatment performance for winery wastewater. *Water Environment Research*, 75 (5), 412–421.

Grismer, M. E., & Collison, R. (2017). The Zeolite-Anammox treatment process for nitrogen removal from wastewater—A review. *Water*, 9 (11), 901. doi:10.3390/w9110901.

Grismer, M. E., Tausendschoen, M., & Shepherd, H. L. (2001). Hydraulic characteristics of a sub-surface flow constructed wetland for winery effluent treatment. *Water Environment Research*, 73 (4), 466–477.

Headley, T. R., Herity, E., & Davison, L. (2005). Treatment at different depths and vertical mixing within a 1-m deep horizontal subsurface-flow wetland. *Ecological Engineering*, 25 (5), 567–582.

Headley, T. R., & Kadlec, R. H. (2007). Conducting hydraulic tracer studies of constructed wetlands: A practical guide. *Ecohydrology & Hydrobiology*, 7 (3–4), 269–282.

Hodgson, C. J., Perkins, J., & Labadz, J. C. (2003). Evaluation of bio-tracers to monitor effluent retention time in constructed wetlands. *Letters in Applied Microbiology*, 36 (6), 362–371.

Hoffmann, H., & Platzer, C. (2010). Constructed wetlands for greywater and domestic wastewater treatment in developing countries. Sustainable Sanitation and Ecosan Program of Deutsche Gesellschaft Für Technische Zusammenarbeit (GTZ) GmbH, Germany.

Hoffmann, H., Platzer, C., Winker, M., & von Muench, E. (2011). Technology review of constructed wetlands: Sub-surface flow constructed wetlands for greywater and domestic wastewater treatment. Deutsche Gesellschaft für Internationale Zusammenarbeit (GIZ) GmbH, Eschborn, Germany, 11.

Ioannidou, V. G., & Pearson, J. M. (2018). Hydraulic and design parameters in full-scale constructed wetlands and treatment units: Six case studies. *Environmental Processes*, 5 (1), 5–22.

IWA (International Water Association Specialist Group on Use of Macrophytes in Water Pollution Control), 2000. Constructed Wetlands for Pollution Control: Processes, Performance, Design and Operation. IWA Publishing, London, UK, 156 pp.

Jabali, M. M., Okhravi, S., Eslamian, S., & Gohari, S. (2017). Water conservation techniques. In *Handbook of Drought and Water Scarcity* (pp. 501–520). CRC Press, New York.

Jenkins, G. A., & Greenway, M. (2007). Restoration of a constructed stormwater wetland to improve its ecological and hydrological performance. *Water Science and Technology*, 56 (11), 109–116.

Jury, W. A., Gardner, W. R., & Gardner, W. H. (1991). *Soil Physics*. John Wiley & Sons, New York.

Jury, W. A., & Sposito, G. (1985). Field calibration and validation of solute transport models for the unsaturated zone. *Soil Science Society of America Journal*, 49 (6), 1331–1341.

Kadaverugu, R. (2016). Modeling of sub-surface horizontal flow constructed wetlands using OpenFOAM. *Modeling Earth Systems and Environment*, 2 (2), 55. doi:10.1007/s40808-016-0111-0.

Kadlec, R. H. (1994). Detention and mixing in free water wetlands. *Ecological Engineering*, 3 (4), 345–380.

Kadlec, R. H. (2000). The inadequacy of first-order treatment wetland models. *Ecological Engineering*, 15 (1–2), 105–119.

Kadlec, R. H. (2007). Tracer and spike tests of constructed wetlands. *Ecohydrology & Hydrobiology*, 7 (3–4), 283–295.

Kadlec, R. H. (2009). Comparison of free water and horizontal subsurface treatment wetlands. *Ecological Engineering*, 35 (2), 159–174.

Kadlec, R. H., Bastiaens, W., & Urban, D. T. (1991). Hydrological design of free water surface treatment wetlands. *Constructed Wetlands for Water Quality Improvement* (pp. 77–86). West Florida University, Pensacola, FL.

Kadlec, R. H., & Knight, R. L. (1996). *Treatment Wetlands*. Lewis Publishers, Boca Raton, FL.

Kadlec, R. H., Knight, R. L. (2006). *Treatment Wetlands*, 1st ed. Lewis Publishers, Boca Raton, FL.

Kadlec, R. H., Tanner, C. C., Hally, V. M., & Gibbs, M. M. (2005). Nitrogen spiraling in sub-surface-flow constructed wetlands: Implications for treatment response. *Ecological Engineering*, 25 (4), 365–381.

Kadlec, R. H., & Wallace, S. (2008). *Treatment Wetlands*. CRC Press, Boca Raton, FL.

Kickuth, R. (1984). The Root Zone Method. Gesamthochschule Kassel-Uni des Landes, Hessen.

Korkusuz, E. A. (2005). Manual of practice on wetlands for wastewater treatment and reuse in Mediterranean countries. Report AVKR, 5.

Langergraber, G. (2008). Modeling of processes in subsurface flow constructed wetlands: A review. *Vadose Zone Journal*, 7 (2), 830–842.

Langergraber, G. (2011). Numerical modelling: A tool for better constructed wetland design? *Water Science and Technology*, 64 (1), 14–21.

Lee, H. Y., & Shih, S. S. (2004). Impacts of vegetation changes on the hydraulic and sediment transport characteristics in Guandu mangrove wetland. *Ecological Engineering*, 23 (2), 85–94.

Levenspiel, O. (1972). *Chemical Reaction Engineering*, 2nd ed. Wiley, New York.

Li, X., Ding, A., Zheng, L., Anderson, B. C., Kong, L., Wu, A., & Xing, L. (2018). Relationship between design parameters and removal efficiency for wetlands in China. *Ecological Engineering*, 123, 135–140.

Liu, L., Li, J., Fan, H., Huang, X., Wei, L., & Liu, C. (2019). Fate of antibiotics from swine wastewater in constructed wetlands with different flow configurations. *International Biodeterioration & Biodegradation*, 140, 119–125.

Lloyd, B. J., Leitner, A. R., Vorkas, C. A., & Guganesharajah, R. K. (2003). Under-performance evaluation and rehabilitation strategy for waste stabilization ponds in Mexico. *Water Science and Technology*, 48 (2), 35–43.

Lucke, T., Walker, C., & Beecham, S. (2019). Experimental designs of field-based constructed floating wetland studies: A review. *Science of the Total Environment*, 660, 199–208.

Małoszewski, P., Wachniew, P., & Czupryński, P. (2006). Hydraulic characteristics of a wastewater treatment pond evaluated through tracer test and multi-flow mathematical approach. *Polish Journal of Environmental Studies*, 15 (1), 105–110.

Marsili-Libelli, S., & Checchi, N. (2005). Identification of dynamic models for horizontal sub-surface constructed wetlands. *Ecological Modelling*, 187 (2–3), 201–218.

Martinez, C. J., & Wise, W. R. (2003). Analysis of constructed treatment wetland hydraulics with the transient storage model OTIS. *Ecological Engineering*, 20 (3), 211–222.

Naipal, E., van der Maas, P., Langenhoff, A., & Rijnaarts, H. (2014). Critical design parameters of sub-surface flow constructed wetlands for the removal of organic micro pollutants from wastewater. Wageningen University, Wageningen, the Netherlands.

Naurath, L., Weidner, C., Rüde, T. R., & Banning, A. (2011). A new approach to quantify Na-fluorescein (Uranine) in acid mine waters. *Mine Water and the Environment*, 30 (3), 231–236.

Okhravi, S., Eslamian, S., & Adamowski, J. (2016). Water reuse in rainwater harvesting. In *Urban Water Reuse Handbook*, edited by S. Eslamian (pp. 823–840). CRC Press, Boca Raton, FL.

Okhravi, S., Eslamian, S., & Fathianpour, N. (2017). Assessing the effects of flow distribution on the internal hydraulic behaviour of a constructed horizontal sub-surface flow wetland using a numerical model and a tracer study. *Ecohydrology & Hydrobiology*, 17 (4), 264–273.

Persson, J. (2000). The hydraulic performance of ponds of various layouts. *Urban Water*, 2 (3), 243–250.

Persson, J., Somes, N. L. G., & Wong, T. H. F. (1999). Hydraulics efficiency of constructed wetlands and ponds. *Water Science and Technology*, 40 (3), 291–300.

Persson, J., & Wittgren, H. B. (2003). How hydrological and hydraulic conditions affect performance of ponds. *Ecological Engineering*, 21 (4–5), 259–269.

Philippi, L. S., Sezerino, P. H., Bento, A. P., & Magri, M. E. (2006). Vertical flow constructed wetlands for nitrification of anaerobic pond effluent in southern Brazil under different loading rates. In *10th International Conference on Wetland System for Water Pollution Control* (pp. 23–29). IWA, Lisbioa.

Rajabzadeh, A. R., Legge, R. L., & Weber, K. P. (2015). Multiphysics modeling of flow dynamics, biofilm development and wastewater treatment in a sub-surface vertical flow constructed wetland mesocosm. *Ecological Engineering*, 74, 107–116.

Ramond, J. B., Welz, P. J., Cowan, D. A., & Burton, S. G. (2012). Microbial community structure stability, a key parameter in monitoring the development of constructed wetland mesocosms during start-up. *Research in Microbiology*, 163 (1), 28–35.

Reed, S. C. (1990). *Natural Systems for Wastewater Treatment* (Manual of Practice, Fd-16). Water Pollution Control Federation, Alexandria, VA.

Reed, S. C., Crites, R. W., & Middlebrooks, E. J. (1995). *Natural Systems for Waste Management and Treatment*, 2nd ed. McGraw-Hill, New York.

Ren, Y., Zhang, B., Liu, Z., & Wang, J. (2007). Optimization of four kinds of constructed wetlands substrate combination treating domestic sewage. *Wuhan University Journal of Natural Sciences*, 12 (6), 1136–1142.

Ronkanen, A. K., & Kløve, B. (2007). Use of stabile isotopes and tracers to detect preferential flow patterns in a peatland treating municipal wastewater. *Journal of Hydrology*, 347 (3–4), 418–429.

Ronkanen, A. K., & Kløve, B. (2008). Hydraulics and flow modelling of water treatment wetlands constructed on peatlands in Northern Finland. *Water Research*, 42 (14), 3826–3836.

Rousseau, D. P., Vanrolleghem, P. A., & De Pauw, N. (2004). Constructed wetlands in Flanders: A performance analysis. *Ecological Engineering*, 23 (3), 151–163.

Rousseau, D. P. L., Lesage, E., Story, A., Vanrolleghem, P. A., & De Pauw, N. (2008). Constructed wetlands for water reclamation. *Desalination*, 218 (1–3), 181–189.

Sabatini, D. A. (2000). Sorption and intraparticle diffusion of fluorescent dyes with consolidated aquifer media. *Groundwater*, 38 (5), 651–656.

Safieddine, T. (2009). Hydrodynamics of waste stabilization ponds and aerated lagoons. ProQuest, Ann Arbor, MI.

Samsó, R., & García, J. (2013). Bacteria distribution and dynamics in constructed wetlands based on modelling results. *Science of the Total Environment*, 461, 430–440.

Shepherd, H. L., Grismer, M. E., and Tchobanoglous, G. 2001. Treatment of high-strength winery wastewater using a subsurface flow constructed wetland. *Water Environmenal Research*, 73 (4), 394–403.

Smith, E., Gordon, R., Madani, A., & Stratton, G. (2005). Cold climate hydrological flow characteristics of constructed wetlands. *Canadian Biosystems Engineering*, 47, 1–1.

Steer, D., Fraser, L., Boddy, J., & Seibert, B. (2002). Efficiency of small constructed wetlands for sub-surface treatment of single-family domestic effluent. *Ecological Engineering*, 18 (4), 429–440.

Su, T. M., Yang, S. C., Shih, S. S., & Lee, H. Y. (2009). Optimal design for hydraulic efficiency of free-water-surface constructed wetlands. *Ecological Engineering*, 35 (8), 1200–1207.

Sudarsan, J. S., Roy, R. L., Baskar, G., Deeptha, V. T., & Nithiyanantham, S. (2015). Domestic wastewater treatment performance using constructed wetland. *Sustainable Water Resources Management*, 1 (2), 89–96.

Suliman, F., French, H., Haugen, L. E., Klöve, B., & Jenssen, P. (2005). The effect of the scale of horizontal sub-surface flow constructed wetlands on flow and transport parameters. *Water Science and Technology*, 51 (9), 259–266.

Suliman, F. F. H. K., French, H. K., Haugen, L. E., & Søvik, A. K. (2006). Change in flow and transport patterns in horizontal sub-surface flow constructed wetlands as a result of biological growth. *Ecological Engineering*, 27 (2), 124–133.

Suliman, F., Futsaether, C., & Oxaal, U. (2007). Hydraulic performance of horizontal sub-surface flow constructed wetlands for different strategies of filling the filter medium into the filter basin. *Ecological Engineering*, 29 (1), 45–55.

Sun, H., Hu, Z., Zhang, J., Wu, W., Liang, S., Lu, S., & Liu, H. (2016). Determination of hydraulic flow patterns in wetlands using hydrogen and oxygen isotopes. *Journal of Molecular Liquids*, 223, 775–780.

Thackston, E. L., Shields Jr, F. D., & Schroeder, P. R. (1987). Residence time distributions of shallow basins. *Journal of Environmental Engineering*, 113 (6), 1319–1332.

Tilley, E., Ulrich, L., Luethi, C., Reymond, P., & Zurbruegg, C. (2014). *Compendium of Sanitation Systems and Technologies*, 2nd Rev. ed. Swiss Federal Institute of Aquatic Science and Technology (Eawag), Duebendorf, Switzerland.

UN-HABITAT. (2008). *Constructed Wetlands Manual*. UN-HABITAT Water for Asian Cities Programme, Kathmandu, Nepal.

Vymazal, J. (1996). The use of sub-surface-flow constructed wetlands for wastewater treatment in the Czech Republic. *Ecological Engineering*, 7 (1), 1–14.

Vymazal, J. (2009). The use of wetlands with horizontal sub-surface flow for various types of wastewater. *Ecological Engineering*, 35 (1), 1–17.

Vymazal, J. (2011). Long-term performance of constructed wetlands with horizontal sub-surface flow: Ten case studies from the Czech Republic. *Ecological Engineering*, 37 (1), 54–63.

Vymazal, J. (2019). Constructed wetlands for wastewater treatment. *Encyclopedia of Ecology, 2nd Edition*, 1, 14–21.

Walker, C., Tondera, K., & Lucke, T. (2017). Stormwater treatment evaluation of a constructed floating wetland after two years operation in an urban catchment. *Sustainability*, 9 (10), 1687. doi:10.3390/su9101687.

Wallace, S. D., & Knight, R. L. (2006). *Small-scale Constructed Wetland Treatment Systems: Feasibility, Design Criteria and O & M Requirements*. IWA Publishing, London, UK.

Wang, B., Jin, M., Nimmo, J. R., Yang, L., & Wang, W. (2008). Estimating groundwater recharge in Hebei Plain, China under varying land use practices using tritium and bromide tracers. *Journal of Hydrology*, 356 (1–2), 209–222.

Wang, R., Li, R., Li, J., & Hu, C. (2014a). A hydraulics-based analytical method for artificial water replenishment in wetlands by reservoir operation. *Ecological Engineering*, 62, 71–76.

Wang, Y., Song, X., Liao, W., Niu, R., Wang, W., Ding, Y., Wang, Y., & Yan, D. (2014b). Impacts of inlet-outlet configuration, flow rate and filter size on hydraulic behaviour of quasi-2-dimensional horizontal constructed wetland: NaCl and dye tracer test. *Ecological Engineering*, 69, 177–185.

Werner, T. M., & Kadlec, R. H. (1996). Application of residence time distributions to stormwater treatment systems. *Ecological Engineering*, 7 (3), 213–234.

Werner, T. M., & Kadlec, R. H. (2000). Wetland residence time distribution modeling. *Ecological Engineering*, 15 (1–2), 77–90.

Xavier, M. L. M., Janzen, J. G., & Nepf, H. (2018). Numerical modeling study to compare the nutrient removal potential of different floating treatment island configurations in a stormwater pond. *Ecological Engineering*, 111, 78–84.

Yousefi, Z., & Mohseni-Bandpei, A. (2010). Nitrogen and phosphorus removal from wastewater by sub-surface wetlands planted with Iris pseudacorus. *Ecological Engineering*, 36 (6), 777–782.

Zahraeifard, V., & Deng, Z. (2011). Hydraulic residence time computation for constructed wetland design. *Ecological Engineering*, 37 (12), 2087–2091.

Zhang, D., Gersberg, R. M., & Keat, T. S. (2009). Constructed wetlands in China. *Ecological Engineering*, 35 (10), 1367–1378.

Zhi, W., & Ji, G. (2012). Constructed wetlands, 1991–2011: A review of research development, current trends, and future directions. *Science of the Total Environment*, 441, 19–27.

Printed in the United States
by Baker & Taylor Publisher Services